The Fight Against Food Shortages and Surpluses

Perspectives of a Practitioner

Front cover: Crowd of protestors in Tunisia in 2011 demanding Aīsh, huriyya, 'adāla igtimā'iyya (bread, freedom, social justice). Of these three, bread is the priority. The price of bread on the world market had escalated; it had become unaffordable for millions of families in the world. The shortage of grain was mainly due to bad government policies in both rich and poor countries, low world stocks and poor weather. In the Arab world, the high price of bread was the trigger of mass protests and years of political turmoil.

Photo: Christophe Ena, AP/Shutterstock

The Fight Against Food Shortages and Surpluses

Perspectives of a Practitioner

John McClintock

CABI

CABI is a trading name of CAB International

CABI
Nosworthy Way
Wallingford
Oxfordshire OX10 8DE
UK

Tel: +44 (0)1491 832111
Fax: +44 (0)1491 833508
E-mail: info@cabi.org
Website: www.cabi.org

CABI
WeWork
One Lincoln St
24th Floor
Boston, MA 02111
USA
T: +1 (617)682-9015
E-mail: cabi-nao@cabi.org

A catalogue record for this book is available from the British Library, London.
References to internet websites (URLs) were accurate at the time of writing.

Library of Congress Cataloging-in-Publication Data

Names: McClintock, John, author.
Title: The fight against food shortages and surpluses : perspectives of a practitioner / John McClintock.
Description: Boston, MA : CAB International, [2020] | Includes bibliographical references and index. | Summary: "This book presents a logical argument, rigorously formulated, for practical policy options to fight food shortages and surpluses. It navigates the reader through issues and ideologies surrounding hunger, presents and tests hypotheses with theory and evidence, and concludes with workable policies to address food crises"– Provided by publisher.
Identifiers: LCCN 2020018098 (print) | LCCN 2020018099 (ebook) | ISBN 9781786394842 (hardback) | ISBN 9781786394859 (ebook) | ISBN 9781786394866 (epub)
Subjects: LCSH: Grain–Prices. | Grain trade. | Food security.
Classification: LCC HD9000.4 .M335 2020 (print) | LCC HD9000.4 (ebook) | DDC 331.1/331–dc23
LC record available at https://lccn.loc.gov/2020018098
LC ebook record available at https://lccn.loc.gov/2020018099

ISBN: 9781786394842 (hardback)
9781786394859 (e-pdf)
9781786394866 (e-pub)

Commissioning Editor: Rachael Russell
Editorial Assistant: Emma McCann
Production Editor: Shankari Wilford

Typeset by Exeter Premedia Services Pvt Ltd, Chennai, India
Printed and bound in the UK by CPI Group (UK) Ltd, Croydon, CR0 4YY

Contents

About the Author vii

Foreword ix

Preface xi

Summary xiii

1 Introduction 1

2 Grain in the World 9

3 The Pattern of Grain Prices 16

4 The Root Cause of Unstable Grain Prices 40

5 Unstable Grain Prices – Who Wins? Who Loses? 63

6 Stabilizing Grain Prices in a Closed Economy 81

7 A Model of the International Grain Market 100

8 Stabilizing Grain Prices in an Open Economy 119

9 Buffer Stocks – Technical and Legal Aspects 140

10 Attitudes towards Price Stabilization and Buffer Stocks 160

11 Conclusion 169

Appendix: The International Wheat Agreement – A Case Study 173

Index 177

About the Author

John McClintock is an agricultural economist (Reading and Oxford universities, École Nationale Supérieure Agronomique, Montpellier). He is an Irish citizen and has worked in a number of developing and European countries and for the European Commission. He is the founder of ACTION for Food Reserves, a non-governmental organization based in Brussels that seeks to promote stable grain prices and the use of buffer stocks.

By the same author:

The Uniting of Nations: An Essay on Global Governance (3rd edition). Published by PIE Peter Lang, Brussels, 2010

An explanation of the United Nations, the European Union and how the limited and partial sharing of sovereignty can make it easier for governments to resolve global challenges by creating international law that is both fair and enforceable.

Foreword

I have always been a fan of stabilizing agricultural product prices. In a market economy, the function of price movements is to communicate important information between producers and consumers. Perhaps new technology has made farming more productive, or farm input prices have fallen, allowing farmers profitably to supply more at previous prices. Rising consumer incomes or changing dietary preferences may have led consumers to wish to buy more of a food product. But sharp and erratic movements in market prices caused by unplanned changes in production – usually weather- or disease-related – together with the chronic price cycles to which some agricultural product markets are prone, do not communicate any useful message between producers and consumers, other than as a short-term and often painful way of informing consumers that less is available this year, or farmers that more produce has to find a market. Persistent agricultural market price volatility of this kind then obscures the important messages that longer-term changes in market prices should be communicating on the evolution of underlying patterns of consumer food demand and agricultural product supply.

Governments can introduce policies to offset the damaging consequences for food consumers of high prices or for farmers of low prices, but the obscuring nature of price volatility remains. Alternatively, governments can attempt to 'fix' market prices, but when ministers announce that they are to introduce a policy to 'stabilize' food prices, what they really mean is 'stop them going up'. When they say they wish to stabilize farm product prices, they mean 'stop them going down'. Market prices then begin to communicate false messages between farmers and food consumers, leading increasingly to too much or too little being produced.

In this book, John McClintock explores in detail the causes of extreme price volatility in agricultural product markets and assesses the merits of alternative policies to deal with their adverse consequences. He advocates the deployment of public grain stocks to stabilize market prices while retaining the important messages of longer-term price trends. Storing grain between harvest seasons is probably the first example in history of an agricultural policy. However, sometimes the oldest remedies are the best.

Christopher Ritson
Emeritus Professor of Agricultural Marketing
Newcastle University, UK

Preface

I owe an inestimable debt of gratitude to the agricultural economist Christopher Ritson, Emeritus Professor of Agricultural Marketing, Newcastle University, UK. He has generously shared his understanding and deep knowledge of the issues raised in this book. The text has been greatly improved by his assiduous attention to detail and, at the same time, his broad perspective of food and agriculture in the world.

I am also most grateful to Dr Mark Corner of the University of Leuven, Belgium, who has, over the years, provided much encouragement and advice on my academic pursuits in the field of food security.

Thanks are also due to the participants who kindly shared their knowledge with me at an expert meeting on stocks, markets and stability, organized by the Food and Agriculture Organization in Rome in January 2014. I extend my appreciation to Dr David Hallam and his colleagues for their invitation to attend that meeting.

I express my warmest appreciation to all the colleagues and friends with whom I have worked in various tropical countries and at the European Commission. Their insights into agricultural matters have proved invaluable.

I would like to thank the staff of CAB International for their guidance and patience.

Last, but by no means least, I thank my wife and children for tolerating the time that I spent away from them in researching the subject matter of the book. I alone am responsible for any errors and deficiencies in the text.

For those who may wish to pursue the ideas presented in the pages that follow, they are most welcome to contact me at: john.mcclintock@action-for-food-reserves.org

John McClintock
ACTION for Food Reserves (NGO)
Brussels
February 2020

Summary

Chapter 1: The world prices of agricultural commodities are unstable. Prices in a country can rise when food is in short supply. This is usually due to bad weather. Shortages can also arise if the food chain is disrupted by, for instance, an epidemic or a pandemic. The opposite may also happen – a fall in prices. This is due to a surplus of food on the market. International trade helps to moderate price instability. However, international trade cannot play a moderating role if there is a worldwide shortage or surplus. Governments need further measures. Taking the food grains as an example – humanity's most basic food – this book explores what these further measures may be.

Chapter 2: This chapter paints the overall picture of grain production and trade in the world. Of the world's 200 countries, 50 produce more grain than they consume. They export grain. The remaining 150 countries do not grow enough for their needs. They fill the gap with imports.

Chapter 3: This chapter describes the pattern of grain prices on the world market over the last 60 years. Prices have tended to be low for periods of 10–20 years at a time. These periods are interspersed with sudden and unexpected surges in prices. Three factors are closely associated with sudden increases in the world price of grain: poor weather during the growing season, a low level of grain stocks in the world, and governmental trade measures (such as import subsidies and export bans). The pattern of prices may be different in the future; prices may become relatively high for periods at a time, interrupted by short episodes when prices slump.

Chapter 4: The price of food – indeed the price of most things – is determined by supply and demand. This chapter explains what this means. Once we have a good understanding of supply and demand we can begin to unravel the fundamental reasons why food prices are volatile. The supply of food is different to that of many other goods. Once farmers have planted their fields, they have relatively little influence over how much is produced. A lot depends on nature, especially the weather. Also, the demand for food is different to that of many other goods in that food is essential for survival. With these features in mind, this chapter modifies the standard model of the market. The modified model enables us to unravel the reasons why the price of food grains is volatile.

Chapter 5: Price volatility causes harm. When prices are high, some citizens cannot afford to buy enough food for their families. *In extremis*, there can be a famine. When prices are low, farmers can be put out of business and forced to leave the land. In dealing with angry farmers and hungry consumers, governments can be disrupted – even thrown out – by food riots and farmer demonstrations. More fundamentally, unstable agricultural commodity prices can impede the development of farming. In many countries of the world, the lack of progress in farming is holding back economic development.

Chapter 6: What can governments do to keep the market price of grain reasonably stable? This chapter considers the closed economy (no trading with other countries). To investigate how the market works, we develop a model of grain supply and grain demand in a closed economy. We suggest that governments establish a 'price band' and keep the price of grain inside the band. We explain how a buffer stock can enable the government to keep the price of grain on its home market within the price band. We explore the options (consumer food subsidies, rationing and deficiency payments) that are available to governments to mitigate the damage caused by volatile prices. The chapter discusses the pros and cons of fixed prices – that is, prices that are imposed upon the market and which replace equilibrium prices. In contrast to fixed prices, a buffer stock does not impose prices upon the market. Rather, a buffer stock offers to sell grain to consumers to prevent prices rising too high and offers to buy grain from farmers to prevent prices falling too low. We conclude that the most plausible approach is to establish a grain reserve to act as a buffer stock.

Chapter 7: Most countries are open economies, trading agricultural commodities with other countries. We elaborate upon the model of grain supply and grain demand that we used in the previous chapter in order to explain the mechanics of the grain market at the level of the world. The model helps us to explore a wide variety of questions concerning open economies. It reveals how the world price of grain is determined. It also shows how countries that are small in terms of their trade in grain are price takers, while large countries are not.

Chapter 8: This chapter uses a model to examine the options for open economies that wish to stabilize the price of grain on their national markets. In some situations, free trade is the answer to the volatility of grain prices. Free trade can solve a country's shortages or surplus, but only if the world price lies within the county's price band. If not, there are two options: a buffer stock or trade measures. A government can keep the price on its national market inside the price band by buying into a buffer stock and selling out of a buffer stock, as the need arises. Trade measures are subsidies or taxes on imports and exports. The problem with trade measures – apart from their expense to the national budget – is harm to other countries. They destabilize the world market price. In contrast, a buffer stock helps to stabilize the world market price. Trade measures cause collateral damage to the rest of the world while buffer stocks are 'globally friendly'.

Chapter 9: This chapter examines buffer stocks in greater detail. The main cost of a buffer stock is in setting it up in the first place. Once set up, the government buys low and sells high and earns a profit. Thus a buffer stock can be financially self-sustaining. The chapter argues that a buffer stock is perfectly compatible with a market economy and, by eliminating the random element from the market price, brings about a more optimal allocation of resources. This increases economic efficiency. The chapter discusses the mechanics of buffer stocks. It discusses their compatibility with the rules of the World Trade Organization. It argues that governments can save money if, rather than setting up national buffer stocks, they co-operate together and set up a multi-country, regional stock.

Chapter 10: So far, the book has argued that grain buffer stocks are a practical and sensible way for governments to stabilize the price of grain on their markets. So why do few governments set up buffer stocks? This chapter considers the experience of the United Nations in its pursuance of stable commodity prices in the 1950s and 1960s. It emphasizes the dominance of the neoliberal ideology over the past three decades which has caused many academics and governments to abandon the aspiration of stable commodity prices.

Chapter 11: The book concludes that it would be sensible for all countries to equip themselves with a grain buffer stock. It will help to relieve the humanitarian and political fall-out of volatile grain prices. Stable prices are also necessary for farmers to adjust their farming methods to the exigencies of climate change. Finally, getting agricultural prices right is a precondition for the economic development and progress of poor countries, without which it will be difficult for them to play their full role in the solution of global challenges.

1 Introduction

1.1 What Is this Book About?

This book is about two of the most fundamental challenges in the world: how to keep the price of food at a level that people all over the world can afford and how to prevent, in both rich and poor countries, slumps in the price of agricultural commodities, which can threaten the livelihoods of farming families. These are both real, live issues. For instance, in 2007/8, there were food riots in 30 countries of the world, in which people were killed. During the last ten years, in many countries, farmers have been plagued by unpredictable and volatile prices for their products.

This book discusses and explores the stabilization of food commodity prices in order to avoid, on the one hand, sudden increases in the price of food and, on the other, slumps in the price at which farmers sell their products.

Trade helps to avoid food shortages and surpluses. Adam Smith (1723–1790), the Scottish classical economist and philosopher, noted this in his magisterial book *An Inquiry into the Nature and Causes of the Wealth of Nations.* He wrote:

> Were all nations to follow the liberal system of free exportation and free importation, the different states into which a great continent was divided would so far resemble the different provinces of a great empire. As among the different provinces of a great empire the freedom of the inland trade appears, both from reason and experience, not only the best palliative of a dearth but the most effectual preventative of a famine.
>
> (Smith, 2007–348)

Over the last 30 years, the world has taken big steps towards freer trade in agricultural commodities.[1] Adam Smith would have given his approval. But free trade cannot magic up commodities from thin air when there is a global shortage; nor can free trade spirit commodities away when there is a global glut and oversupply.

This book explores what governments and regional organizations can do to keep the price of agricultural commodities stable on their own national markets if the world price becomes too high or too low. It focuses on one group of agricultural commodities – the food grains. They keep humanity alive. They include wheat, rice, maize and many other staple foods. The conclusions of the book apply, however, to all agricultural commodities.[2]

1.2 Buffer Stocks Can Stabilize Grain Prices – so Why Are They Controversial?

The argument of this book is that a grain reserve used as a buffer stock is a plausible way of bringing greater stability to the price of grain. Some countries, such as China and India, already have a buffer stock. But many countries do not, including the United States and most of the countries of the European Union. Some people believe that a buffer stock is too much interference with the market. Others fear that it will end up as unwanted 'food mountains'. Other people fear that a buffer stock would cost too much. Lastly, some rich governments appear content to buy themselves out of the problem of price volatility by subsidizing imports and exports.

This book refutes these points of view. A buffer stock does not prevent the market from operating efficiently. Prices are allowed to vary according to supply and demand but the variation is kept within limits. As for the fear of 'food mountains' – these arose in the 1970s and 1980s not because the governments were controlling the volatility of grain prices but because they had permanently raised the prices of farm commodities above their long-term equilibrium level. A buffer stock does not raise the price of grain, it simply moderates its volatility. Regarding costs, the main cost is the initial set-up cost. A buffer stock can be, in principle, financially self-sustaining. Finally, when governments extricate themselves from a price problem with subsidies or taxes on imports and exports, they make the situation worse for everybody else in the world. Governments need to act in a manner that is 'globally friendly' not in a way that complicates the lives of others.

Almost as important as food security is energy security. In the 1970s, many rich countries, including the United States and the countries of the European Union decided to increase their energy security. They did it by setting up strategic petroleum reserves to guard against disruptions of the supply of oil. These reserves moderate the volatility of oil prices on a country's market. Just as a strategic petroleum reserve can moderate the volatility of oil prices, so grain buffer stocks can moderate the volatility of food prices.

1.3 Stable Prices at the National Level or at the World Level?

Governments have long wanted commodity prices to be stable. Almost 100 years ago, governments set up the International Wheat Agreement with the express purpose of moderating the volatility of world prices. In the 1950s and 1960s, the United Nations set up schemes to stabilize the world prices of five other commodities – tin, sugar, coffee, cocoa and rubber. These were international agreements between governments. In principle, the agreements were designed to improve the situation for all countries, irrespective of whether they were producers, consumers, importers or exporters.

Some of these agreements were successful, but only for a limited period of time. Sooner or later, they collapsed or lapsed. The world now has no schemes in place to moderate the volatility of commodity prices. The agreements did not take root because governments did not co-operate sufficiently. It proved very difficult to find a way whereby the costs and benefits of stabilization were shared fairly between countries. For instance, in the case of the International Wheat Agreement, Canada and the USA felt that they were shouldering more than their fair share of the cost and received too little of the benefit. This caused them to withdraw from the agreement, which subsequently lapsed.

In any international endeavour, it is a big challenge for governments to agree on how to share the costs and benefits between themselves in a manner that is deemed to be fair. This is one reason why so many international endeavours struggle to survive or fail to reach their ambitions. To be truly successful, countries need to be prepared to share, on a limited basis, some of their national sovereignty. This has been achieved in the EU but not elsewhere. The success of the EU demonstrates that countries can co-operate successfully together if they are prepared to share elements of their national sovereignty.

Commodity prices can be stabilized either at the national level or at the world level. To succeed at the world level, governments would have to be prepared to share some elements of national sovereignty at the level of the world. In the short term this seems unlikely, although sharing sovereignty at the regional level seems a possibility.

Price stabilization is, therefore, more likely to succeed if it starts at the level of the individual country (or regional organization) rather than at the level of the world as a whole. It is far easier to reach agreement – on such issues as price bands and how to run a buffer stock – within a single country than it is within the UN with its 200 countries. Thus the focus of this book is on what individual governments can do to moderate price volatility on their own national markets. Of course, once a country has established its own buffer stock, there is no reason why it should not combine it with those of other countries to create a regional buffer stock. The merging of regional buffer stocks could, eventually, lead to a world buffer stock and the stabilization of world prices, as was the aspiration of the UN in the 1960s and 1970s. But the process is more likely to succeed if it starts with individual governments and national (or regional) buffer stocks.

1.4 The Importance of Getting Prices Right

Having control over the price of grain means more than being able to avert food shortages and food surpluses; it means the government has its hand on a crucial lever in the process of national development. All countries start the process of economic development with most of their population growing their own food on smallholder family farms. For a country to develop, the farm sector, consisting of these smallholder family farms, needs to (i) shed labour to the newly emerging industries in the growing towns and cities; (ii) increase the labour productivity of those farmers who remain in the sector so that one farmer can feed a growing number of non-farmers; and (iii) earn enough money to constitute the market for the products produced by the newly emerging industries.

The farm sector has, therefore, a trinity of tasks: to provide the food, the labour and the market for economic development. It can only fulfil these three roles if grain prices are reasonably stable. If they are not stable, farmers find it impossible to plan ahead. They will be reluctant to invest in their farms, to modernize their systems of production and to increase their labour productivity. Unstable grain prices are, therefore, inimical to the balanced and natural development of a country.

This has been made clear by the economic historian Peter Timmer, of Harvard University. He has examined which countries have developed successfully and which have not (Galtier, 2013, p. 11). Regarding the first country to industrialize – England – Timmer has pointed out that during the era of the Corn Laws from the late 1600s to the early 1800s, England protected its farmers from cheaper imports of wheat, barley and other grains from foreign countries. The Corn Laws protected English farmers from low prices. They also stabilized grain prices within England. The Corn Laws moderated the natural volatility of grain prices. The result was, arguably, to stimulate the world's first agricultural revolution and to provide the food, labour and market for manufactures, which were necessary for the first industrial revolution to happen.[3]

The Corn Laws worked by keeping out cheap grain. This benefitted English landowners and farmers. They received higher prices than would have been the case had England pursued a policy of free trade. However, the corollary was that the price of bread for the people was higher than it would have been under free trade. Eventually, bread became too expensive for ordinary people. As a result, the Corn Laws were repealed in 1846. Thereafter, Britain adopted a policy of free trade in agriculture and began to import cheaper grain from North America,

Australia and the European continent. The cost of bread came down for British citizens.

The repeal of the Corn Laws is perhaps the first example of a government decision being influenced by analysis put forward by an agricultural economist (David Ricardo and the theory of rent). In this book, we shall talk a lot about maximum consumer prices and how important it is for governments to avoid the price of food becoming too expensive for consumers. The Corn Laws stimulated farming in Britain and provided the wherewithal for an industrial revolution. After the Industrial Revolution had taken root, the Corn Laws were rescinded. This sequence of events encapsulates the argument of this book: that it is vital for governments to get grain prices right if their countries are to make economic progress.

During the 18th century, when England was rapidly developing, France fell behind in terms of both rural and urban productivity. The country only began to catch up with England in the latter half of the 19th century. This is when it abandoned its long-time strategy of 'provisioning Paris' as cheaply as possible with food grains from the countryside. The failure of this provisioning policy had been a trigger of the French Revolution in 1789. From the 1850s, the French government began to provide policy and investment support to the smallholder farmers who dominated French farming. The government protected its farmers from cheap imports and low prices.

As for Germany, it developed rapidly under Chancellor Otto von Bismarck (1815–1898). The Chancellor forged a pact between 'steel and rye' – in other words between the newly emerging industrial sectors and smallholder farmers. As part of this pact, the smallholders were protected from low grain prices.

On the other hand, Russia's economy has not developed smoothly and naturally. Its farmers were not given reasonable grain prices, which would have induced them to produce a surplus for sale. Instead, its peasant farmers were forced to deliver their grain surpluses to the state. The purpose was to provide sufficient food to feed the country's strategy of forced and rapid industrialization. The failure to gradually develop a modern agricultural system by stimulating the farmers with attractive prices was one major factor in the ultimate collapse of the Soviet Union in the 1990s.

The importance of getting the grain price right is also shown in the experiences of Asian countries. Here there have been both successes and failures. In Japan, early investments in raising productivity on small farms paid high dividends in feeding a growing population in the towns and cities. The higher productivity of the small farms meant that they could shed people who then moved into the newly emerging factories and workshops where the manufacturing sector was in the process of being born. The government's rice policy provided stable prices to rice farmers and protected them from price slumps. The government's policy regarding agricultural and food prices helped to bring economic development.

In contrast, Thailand, after 1880, fell systematically behind Japan in terms of per capita income. This was partly because the government taxed its farmers. It did not encourage farmers to develop and invest. It did not provide them with good prices for their products. Instead, the government exposed farmers to the full force of volatile prices of rice on the world market.

England, France, Germany and Japan are examples of countries that exercised control over their grain prices. Their farmers could make a profit. This enabled the farmers to improve their farms and to take up new farming methods. Some of the farmers left the countryside for the emerging towns and cities. They produced the new goods which they sold back to the countryside, which could afford to buy them. It could afford to buy them because the farmers were making a reasonable profit from farming. Farming was the engine of economic development. In contrast, Russia and Thailand did not look after their farmers in terms of providing them with attractive and stable prices. Development in these countries was delayed.

It is thus important to get grain prices right and to keep them right. This is not only so that farmers can remain in business and the people can afford to eat; but also so that the farm sector can grow. This is one of the preconditions for national economic development.

1.5 What Is Different about this Book?

This book differs from much of the conventional wisdom on food shortages and hunger in the following ways: It regards hunger as an economic problem. It does not regard hunger as physical problem – as a lack of physical food.

Hunger arises because people cannot afford to pay the price of food. In this respect the book belongs to the school pioneered by Amartya Sen in the early 1980s. He argued against the simplistic 'food availability decline' explanation of hunger and famine (Sen, 1981). Famines can happen irrespective of whether there is more food, less food or the same quantity of food in the world. What matters is whether people can afford to pay the price of food and whether they have physical access to it, i.e. whether the people and the food are in the same location.

The book has a wider scope than many conventional analyses of hunger and famine. These tragedies are but one side of a broader coin. The other side of the coin is surpluses of food commodities. The name of the coin is food price volatility. This is the focus of the book. Food price volatility gives rise to the opposing tragedies of, on the one hand, hunger and, on the other hand, farmers being driven off the land due to rock-bottom prices. Both are symptoms of the deeper problem of volatility. We can only avoid these twin tragedies if we can fix the problem of unstable grain prices. The book therefore looks at a context that is broader than some analyses of hunger. These tend to focus on the physical quantity of food that is produced by farmers without always considering whether people can afford to buy what the farmers produce.

On a point of terminology, the book eschews the term 'food security'. This is a term that defies a clear and meaningful definition. For instance, after World War II, food security referred to the capacity of countries to grow enough food to feed their populations. At that time, the term was strongly associated with the supply of food. In the 1980s, the meaning changed and the term became more associated with the demand for food – whether people were able to acquire enough food for a healthy life. Since around 2005, the term has been associated with the concept of sustainable food systems. As well as defying a clear definition, the term evades quantification. It is impossible to state that such-and-such country is, for instance, '75% food secure'. For these reasons, the term is not useful for a book that seeks to base its analysis on facts and figures.

Although the book applies economic market analysis to the problem of price instability, it has been written in a way that makes it accessible to anyone with an interest in the subject.

1.6 Conclusion

For humanitarian, political and strategic reasons, it is important that the prices of agricultural commodities are stable, rather than volatile. The purpose of this book is to explore how governments can keep the price of grain within a national price band.

Notes

[1] The liberalization of world agricultural trade was the aim of the negotiations between the members of the General Agreement on Tariffs and Trade (which subsequently became the World Trade Organization) during the Uruguay round. This round of negotiations was completed in 1994. Volatile prices of agricultural commodities had been a matter of concern during previous decades. It was considered that liberalization would bring about more stable prices on the world market. The reasons why the policies of the time were considered to exacerbate the instability of the world market price are given in Harris *et al.* (1983, p. 258ff) and stem from the fact that rich countries had tended to insulate themselves from the world market, stabilizing the prices on their own home markets by using the world market as a residual market. Rich countries achieved stable prices at home but at the cost of exacerbating the instability of the world market. In their simulation of the benefits of trade liberalization, Tyers and Anderson (1992, p. 226) estimated that the volatility of wheat and rice prices could fall by a quarter (see their Table 6.14). They write: 'The results in Table 6.14 clearly show that world food markets would be very much less volatile if agricultural trade policies were liberalized.' Agricultural policies were, indeed, liberalized in the Uruguay round.

But in 2007/8 and 2011, the prices of wheat and rice became very volatile. This poses some important questions: has liberalization, in fact, calmed the world market and resulted in more stable prices? If there is no specific mechanism put in place to stabilize prices (such as a buffer stock), can free trade ever be a solution to volatile commodity prices?

[2] In contrast to grain, which is produced by most countries mainly for home consumption, there is a number of commodities that are produced mainly for export – such as coffee, cacao and rubber. The findings of this book apply to these commodities as much as to grain. The main difference is that consumers of coffee, cacao, rubber, etc. are in foreign countries, not in the country where these crops are grown.

[3] It is perhaps no coincidence that the first country to manage its farm prices (England) was the first country to industrialize.

References

Galtier, F. (2013) *Managing Food Price Instability in Developing Countries: A Critical Analysis of Strategies and Instruments.* CIRAD, Agence Française de Développement, series A Savoir, Paris.

Harris, S., Swinbank, A. and Wilkinson, G. (1983) *The Food and Farm Policies of the European Community.* John Wiley, Chichester, UK.

Sen, A. (1981) *Poverty and Famines: An Essay on Entitlement and Deprivation.* Oxford University Press, Oxford, UK.

Smith, A. (2007) *An Inquiry into the Nature and Causes of the Wealth of Nations.* Originally published in 1776. Harrison House, Petersfield, UK.

Tyers, R. and Anderson, K. (1992) *Disarray in World Food Markets: A Quantitative Assessment.* Cambridge University Press, Cambridge, UK.

Grain in the World

2

2.1 Introduction

The food grains keep us alive. They provide our daily bread, rice, chipatis (in India), nsima, ugali and mealie meal (in Southern Africa), njera (in Ethiopia) and a host of other staple foods that the world's population eats every day. Although the book tends to focus on the three most important grains – wheat, maize (corn) and rice – the discussion applies equally to the many other grains that provide food. These include teff, sorghum, millet, barley, rye, triticale, oats and fonio. As well as keeping humanity alive, some grains, such as barley and maize, feed farm livestock. A great proportion of humanity's meat, milk, cheese, yogurt and eggs are ultimately derived from grain. In addition, biofuels can be made from food grains.

Table 2.1 shows the level of production of the various food grains.

2.2 Where Is Grain Cultivated?

Figures 2.1–2.3 show where the three most important grain crops – maize (maize), rice and wheat – are grown.

2.3 The Distribution of the World's Population

Figure 2.4 shows where humanity lives. More than half lives in Asia. Generally speaking, the Americas and Australia are rather sparsely populated.

2.4 The Distribution of the World's Cropland

Figure 2.5 shows the distribution of cropland in the world. Most is in Asia – this should not surprise us since Asia is by far the largest continent in terms of area. Europe ranks second. Although Europe is a relatively small continent, its high population density means that most land is cultivated.

Table 2.1. Food grains: their importance and uses.[1]

Grain	Worldwide production (million metric tonnes)	Uses and production area
Maize (corn)	1000	A staple food of people in the Americas and Africa; called corn in North America, Australia and New Zealand. A large portion of maize is grown for livestock feed. Some is used for biofuels.
Rice	750	The primary food of tropical and some temperate regions. Staple food in South Asia, the Far East, Brazil and parts of Africa.
Wheat	750	The primary cereal of temperate regions. It has a worldwide consumption but it is a staple food of North America, Europe, Australia, New Zealand, most of the Southern Cone and much of the Middle East.
Barley	150	Grown on land that is too poor or too cold for wheat. Used for malt (an ingredient in some beverages and foodstuffs) and for livestock feed.
Sorghum	60	Important staple food in Asia and Africa and used as feed for livestock.
Millet	30	An important staple food in Asia and Africa.
Oats	25	For humans, oats are a popular breakfast food. For equines, oats are important as a source of energy. Horses, mules and donkeys are used in many countries for farm work. As recently as the 1950s, half of all farm power in the UK was provided by horses. Consequently, oats was the principal grain at that time.
Rye	15	Important in cold climates where it is used for making bread.
Triticale	15	Hybrid of wheat and rye, grown in cold climates.

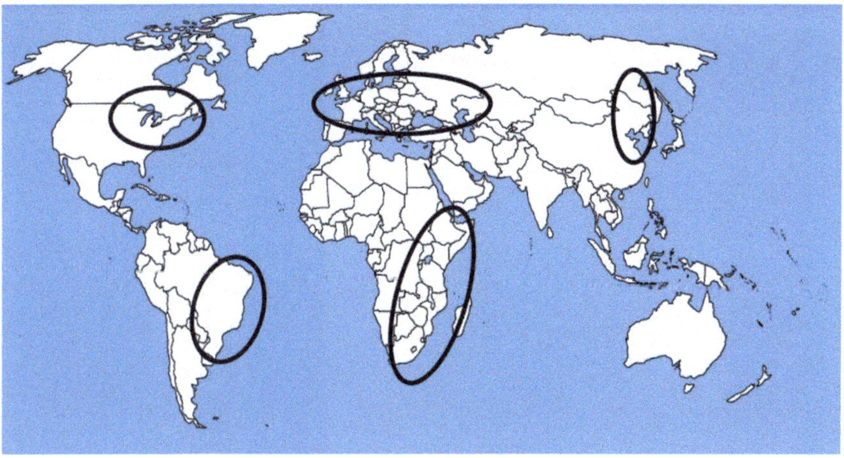

Fig. 2.1. The distribution of maize (corn) production.

2.5 Surplus and Deficit Continents

Sixty per cent of the world's population lives in Asia; yet this continent has only 40% of the world's farmland. Not surprisingly, Asia needs to import grain from other parts of the world. As a whole, and speaking somewhat generally, Asia is a food deficit continent.

Those continents, which were settled by Europeans over the last 200–300 years, have large areas of cultivated land relative to their populations. With much land and a small population, we find that North America, South America and Australia produce more food than they eat. They are food surplus continents.

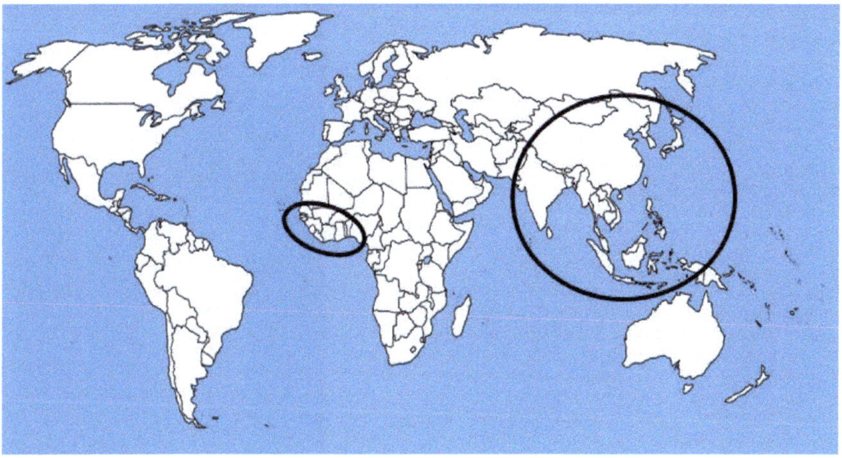

Fig. 2.2. The distribution of rice production.

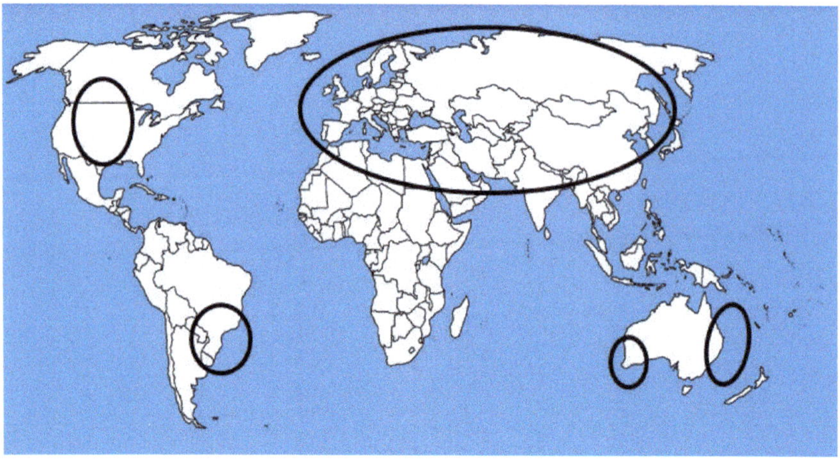

Fig. 2.3. The distribution of wheat production.

Europe is also a food surplus continent. Much of its land is cultivated and its farms are some of the most productive in the world (due to fertile soils, intensive cultivation and a high level of investment in farming since World War II). Europe tends to export its surplus to Africa. At first sight this is somewhat paradoxical; Africa has 16% of the world's population and 17% of the world's cropland. Ostensibly, Africa should not need to import food because it has sufficient cropland for its population. Nevertheless, Africa does import food and much of its imports emanate from Europe.

Several reasons explain this paradox: African agriculture is relatively poorly developed and its crop yields are low; African urban populations

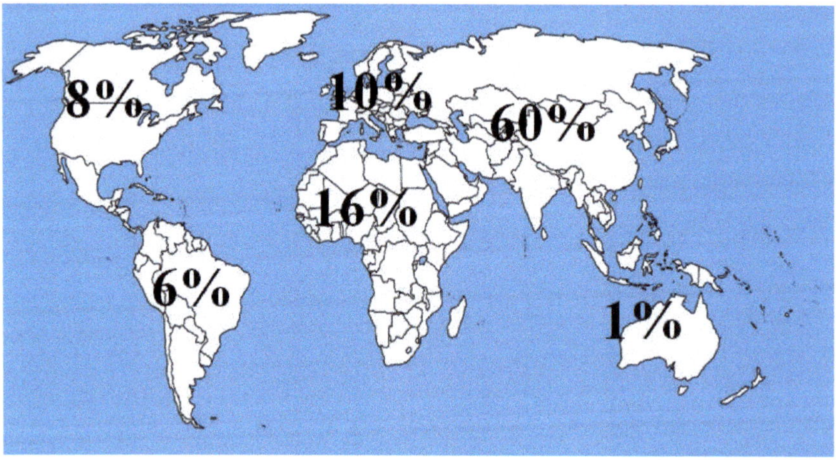

Fig. 2.4. The distribution of the world's population.

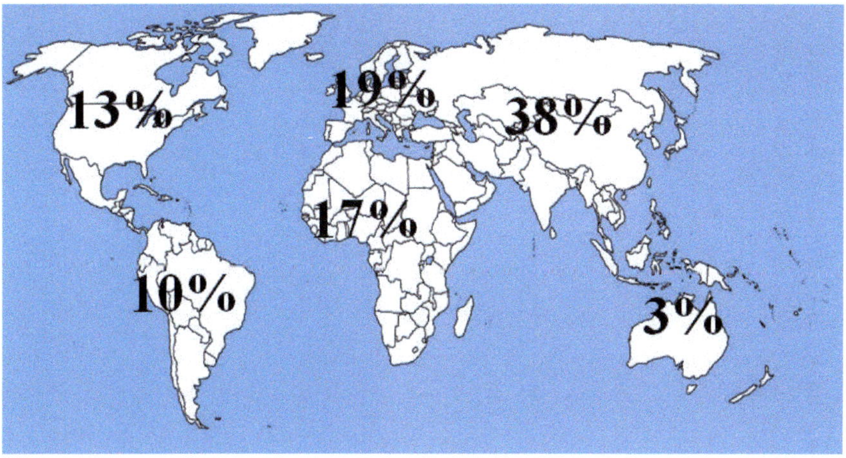

Fig. 2.5. The distribution of the world's cropland.

have grown used to European diets; and Europe has, at least in the past, subsidized some of its food exports to Africa.

2.6 The Grain Trade

Almost all countries of the world either import or export grain. Some countries do both – by, say, importing maize and exporting wheat or by importing one type of wheat and exporting another type. According to the Food and Agriculture Organization of the United Nations, around 400 million tonnes of grain were traded between the countries of the world in 2017. The top 15 exporting countries are shown in Table 2.2 while the top 15 importing countries are shown in Table 2.3.

The two tables below show that the European Union, the USA and the Russian Federation are the world's most important exporters of grain. As for imports of grain, the three biggest importers are the European Union, China and Japan. For every one country that exports grain, there are three that import it. It means that three quarters (150) of the countries of the world rely on imports of grain to feed their populations. This shows the great importance of the world's grain trade in keeping humanity alive.

Although it omits some significant trade flows, Fig. 2.6 shows the overall general pattern of grain trade in the world. Grain flows from the Americas and Australia to the grain-deficit, highly populated continent of Asia. (The diagram is stylized in that grain from the Americas may flow through the Panama Canal or through the Red Sea as well as around Africa on its way to Asia.) Grain also flows from Europe to Africa.

Table 2.2. Top 15 countries that export grain (% of world exports)

European Union	24%
USA	22%
Russian Federation	11%
Ukraine	10%
Argentina	10%
Australia	8%
Brazil	8%
Canada	7%
India	3%
Thailand	3%
Kazakhstan	2%
Vietnam	2%
Turkey	1%
Pakistan	1%
Paraguay	1%

Table 2.3. Top 15 countries that import grain (% of world imports)

European Union	26%
China	9%
Japan	6%
Mexico	6%
Egypt	5%
Saudi Arabia	5%
Vietnam	4%
Indonesia	3%
Bangladesh	3%
Brazil	3%
Turkey	2%
Colombia	2%
Philippines	2%
Nigeria	2%
Malaysia	2%

The European Union is the world's most important exporter of grain. Around half is wheat, the rest being maize and barley in about equal amounts. As for imports of grains, the European Union again ranks as the world's number one. Around half of its imports are wheat, with a third being maize and a tenth being barley.

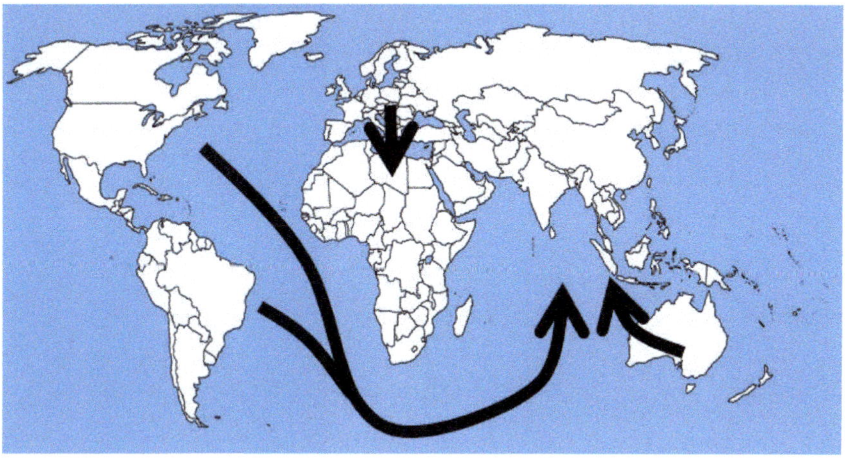

Fig. 2.6. The world's grain trade.

2.7 Summary

Grain provides a major part of humanity's food. There are ten prominent types of grain of which maize (corn), rice and wheat are the most important. The countries of the New World, having relatively low populations and extensive areas of cropland, export grain to the more densely populated continent of Asia. The following chapter turns to the issue of the price of grain.

Note

[1] The table has been adapted from https://en.wikipedia.org/wiki/Cereal (accessed 29 March 2019; updated with production data from FAOSTAT for 2015–2017).

The Pattern of Grain Prices 3

3.1 Introduction

The previous chapter drew the overall picture of grain production and trade. This chapter focuses on the price of grain. It starts with an examination of international food prices in general. It then discusses the pattern of prices of the three most important food grains: maize, wheat and rice. It then focuses on one particular period – that of 1996 to 2008. It attempts to identify the many and various factors that were at play during this period and which are felt to have influenced the price of grain. It was at the end of this period that prices reached a very high level, so high that there were food riots in some 30 countries around the world. In some cases the riots were violent and people were killed.

The chapter concludes by considering the possible impact of climate change on agricultural production and the long-term forces that influence the world prices of food commodities. It asks whether these long-term forces may have fundamentally changed in recent years, leading the world into an era of prices that are higher than during the past 60 years.

3.2 The FAO Food Price Index

The Food and Agriculture Organization (FAO) of the United Nations has calculated a 'food price index' since the early 1960s. It is a composite price of five commodities: cereals (grains), meat, milk, oilseeds and sugar. The index measures the price of these five commodities when they are traded on the international market. It is shown in Fig. 3.1.[1]

Figure 3.1. shows us how the real price of internationally traded food commodities has changed during the last 50 years. Throughout the 1960s, prices remained relatively stable. Since the beginning of the 1970s, the index has become unstable. There was a dramatic rise in prices in the early 1970s. Next, prices proceeded to gradually fall until the 1990s. Following this steady fall, food commodity prices rose, with the rise accelerating from

Fig. 3.1. FAO Food Price Index Deflated, 2002–2004 = 100. (World Food Situation Database, FAO)

around 2005. Then the world witnessed two sudden increases in price: the first in 2007/8 and the second in 2011. The general pattern is one of a cycle – after a sudden rise and fall in the early 1970s, prices fell slowly before gradually rising and again reaching a peak. However, whereas food prices fell dramatically following the peak in the early 1970s, now they are still about 50% above the average level pre-2008.

3.3 The Price of Grain from the 1990s

The FAO provides data from the 1990s for each individual component of its food price index. Figure 3.2 shows the price of grains.

The figure shows the volatility of grain prices. It shows how prices jumped up in the mid-1990s and did so again, to a much larger extent, in 2008/9 and 2011. Since then grain prices have fallen back somewhat, but they remain well above the levels of the 1990s.

3.4 The Price of Individual Grains from the 1960s

3.4.1 During the 1960s, prices were relatively stable

We now focus on three specific grains: maize (known as corn in the USA), wheat and rice. In the three figures below, we do not show real prices as we did in Fig. 3.1.[2] Rather we show nominal prices. Nominal prices are the

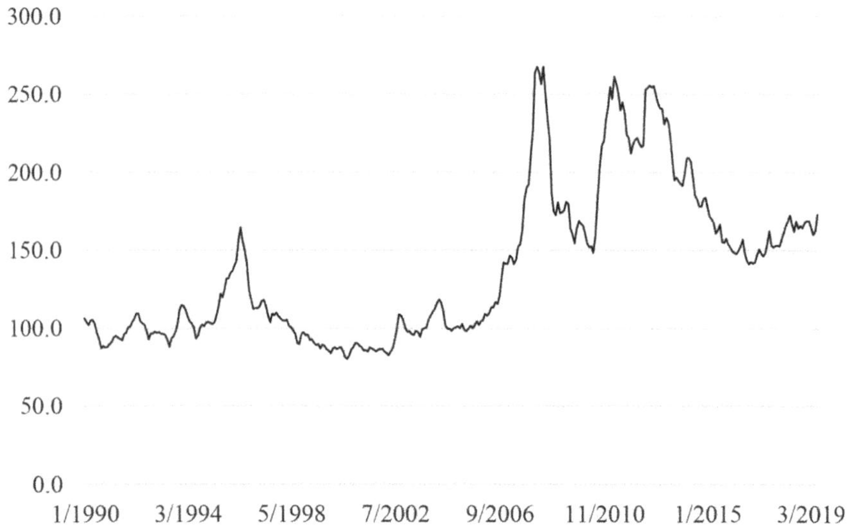

Fig. 3.2. FAO cereals price index monthly – not deflated, starting 1/1990. World Food Situation database, FAO.

prices that are experienced by people in their daily lives and to which they react as parents trying to feed their families and as farmers trying to run an agricultural business. A sharp rise in the nominal price for consumers has a political effect; equally, a sharp fall in the nominal price for farmers has a political effect.

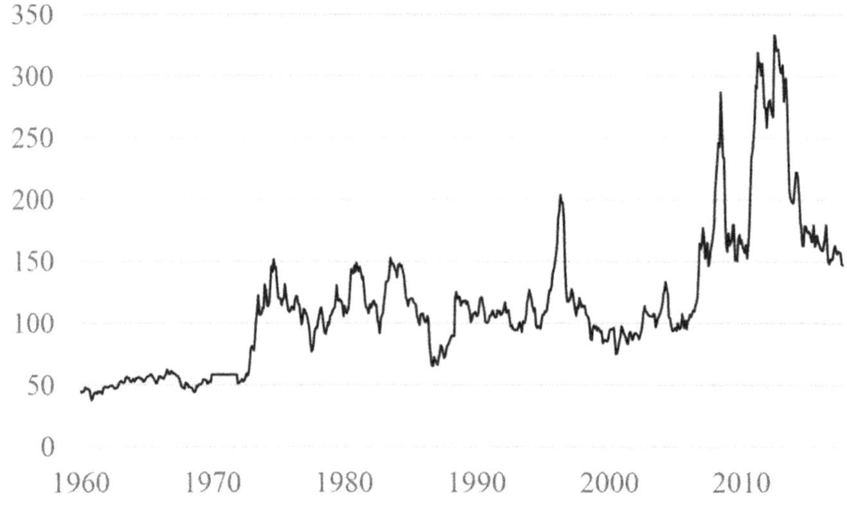

Fig. 3.3. Maize prices since 1960 US $ per metric tonne Nominal. World Bank commodity price data ('the pink sheet')

Fig. 3.4. Wheat prices since 1960; (US$/metric tonne nominal). World Bank commodity price data ('the pink sheet')

Figure 3.3 shows the price of maize on the world market since 1960.[3] Figure 3.4 shows the price of wheat on the world market since 1960. Figure 3.5 shows the price of rice on the world market since 1960.[4]

These three figures reveal that grain prices are unpredictable and volatile. They can rise very quickly and by substantial amounts. It is not uncommon for prices to double or triple within a period of weeks. Having suddenly risen, they can quickly fall back to their original level.

Fig. 3.5. Rice prices since 1960 (US$/metric tonne nominal). World Bank commodity price data ('the pink sheet')

For all three grains, prices were relatively stable up to the early 1970s. This was the result of a deliberate attempt to stabilize prices. In the aftermath of World War II, there was a general wish for a stable and ordered world. This was a natural reaction to the chaos of the war. Moreover, it was felt by statesmen and intellectuals that one of the reasons that the world had descended into war was that the international economy had been in disarray during the 1920s and 1930s. As the war was coming to a close, the victorious countries, led by the USA, were keen to put in place arrangements and institutions that would take the world into a new era of stability. In the area of agriculture, they revived the International Wheat Agreement to moderate the volatility of wheat prices.

Thus it was that the United Nations (UN), the International Monetary Fund (IMF), the World Bank and the Organisation for Economic Co-operation and Development (OECD) were set up in 1944 or soon afterwards. Practically all governments soon joined these organizations as they were seen as paths to prosperity and stability.

This spirit of co-operation extended to international trade in food commodities. The governments of the big importers and big exporters came to an agreement to stabilize the price of wheat. This agreement started in 1949 and lasted until 1971.[5]

The International Wheat Agreement achieved a high level of success. In the 1950s, prices never fluctuated by more than 10% from the previous year. Compared to the 50 years before and the 50 years since, this was remarkable and led Tom Sewell, a staff member of the International Wheat Council, to observe (Sewell, 1992):

> To a newly arrived trader on the present scene it must seem astonishing to reflect that between 1954/5 and 1968/9 the price of US No 2 Hard Red Winter Wheat for fob Gulf ports remained throughout those years between $58 and $66 a ton – that is to say, a range of $8 per tonne over fourteen years.

3.4.2 The world food crisis of the 1970s

In the early 1970s, the stability ended. The prices of many food commodities rose dramatically. This is clearly illustrated by all five figures above, which show the explosion in food prices in the early 1970s. Many governments were caught off-balance with their populations struggling to feed themselves. The UN convened a World Food Conference in 1974 in an attempt to address the problem.

Academics and observers quickly took up their pens to analyse what had happened and why it had happened. There was a profusion of diagnoses. A consensus emerged that two factors had been critical. Firstly, the 1971 and 1972 harvests in the Soviet Union had failed. This necessitated the purchase by the Soviet Union of over a quarter of the USA wheat crop, along with soybeans, corn and other feed grains. This was not only the first-ever major purchase of American grain by the Soviet Union, but it was also the biggest single purchase of American grain in history and the world's largest commercial transaction.

The Soviet Union purchased the grain from half a dozen American grain traders, discreetly negotiating contracts with each individual grain trader. The contracts were negotiated privately, beyond the glare of publicity and without the full awareness of the US government. This meant that the arrival of the Soviet Union on the US grain market passed largely unnoticed. The market price did not rise in anticipation of the prospective deals. The Soviets were therefore able to conclude the contracts and buy the grain at relatively low prices.

After the purchases had become known to the market and the public, the market price of American grain escalated sharply. Consumers in the USA had now to pay more for grain. Inflation and the cost of living rose. The affair was dubbed 'The Great Grain Robbery'. It engendered much heated debate in Congress as to how it had come to pass that the Americans had sold so much grain at such low prices to their Communist opponent (Trager, 1973).

The second cause of the dramatic rise in the price of grain in the 1970s was the oil crisis of the early 1970s. This began in October 1973 when the members of the Organization of the Petroleum Exporting Countries (OPEC) proclaimed an oil embargo. The embargo was targeted at those countries that had supported Israel during the Yom Kippur War. Initially, the targeted countries were Canada, Japan, the Netherlands, the UK and the USA, with the embargo also later extended to Portugal, Rhodesia and South Africa.

By the end of the embargo, in March 1974, the price of oil had risen from US$3 per barrel to nearly $12. The oil crisis had ramifications all around the world and affected all economic sectors. It impacted upon farming because farmers relied upon oil directly for their fuel and indirectly for fertilizers and pesticides. The higher costs of these inputs were passed on to the consumer in the form of higher prices for agricultural commodities.

The increase in cereal prices placed stresses and strains on the International Wheat Agreement. The costs of stabilization increased and governments complained that the costs and benefits of the agreement were not being shared fairly between them. The agreement consequently broke down.

Christopher Ritson of Newcastle University subsequently observed (Ritson, 1977, p. 342):

> Although wheat prices remained fairly stable between 1954 and 1971 this can be attributed largely to a willingness on the part of Canada and the USA to allow their own stocks to accumulate, to a restriction of their wheat acreages, and to the donation by the USA of large quantities of the commodity as food aid. The declaration by the US government that it was no longer prepared

to carry the full cost of storing grain for the world, plus a period of high agricultural commodity prices in 1972–73, increased the general awareness that a successful international grain stabilising programme must contain an agreement about stockholding as well as an agreement about prices.

3.4.3 The period from the mid-1970s to the present

Since the world food crisis of the mid-1970s, grain prices have not been stable. There have been further large fluctuations.

It is possible to discern a rough pattern and a cycle of sudden increases and slumps. The general pattern is that, over several years, grain prices rise slowly and gradually. They then tend to increase more rapidly. At a certain point, they suddenly escalate to a high point. It is at this point that governments become concerned because the high prices of grain at the farm gate are transmitted down the food chain to the prices that consumers have to pay for their bread, tortillas and rice in local markets and shops.

Having reached very high levels, the markets adjust and prices tend to fall as rapidly as they rose. They then descend gradually to a low point. At this point, farming may no longer be profitable. Farmers can face financial losses and bankruptcy. They either cut back on their sown area or leave farming altogether simply because they cannot make enough money.

The rough pattern of grain prices up to the time of writing (2020) has been one of rapid increases and sudden falls followed by slow declines to a low point, with cycles occurring at intervals of between five to 25 years.

3.5 An Analysis of the Price Cycle of 1996–2008

Each time grain prices unexpectedly and suddenly increase there is great concern and even panic, lest the world runs out of food. The spectre of Thomas Malthus, the English parson who predicted in 1798 that the world would run out of food, reappears in books and articles. In newspapers throughout the world, numerous pundits offer instantaneous theories about the causes and are ever prone to find culprits and villains. International conferences are hastily convened to ponder the reasons and come up with remedies for the high price of food.

The more seasoned observers of these grain price cycles have explained them in terms of long-term background conditions and short-term shocks and anomalies. These act on both the supply side and on the demand side. There are many such factors, as an examination of one such price cycle – that of 1996–2008 – reveals.[6]

At this point we draw on analysis performed by the Department of Agriculture of the USA. The department aggregated the prices of four grain commodities – maize, rice, soyabeans and wheat. Rice and wheat are used mainly as human food, maize is used as both food for humans and

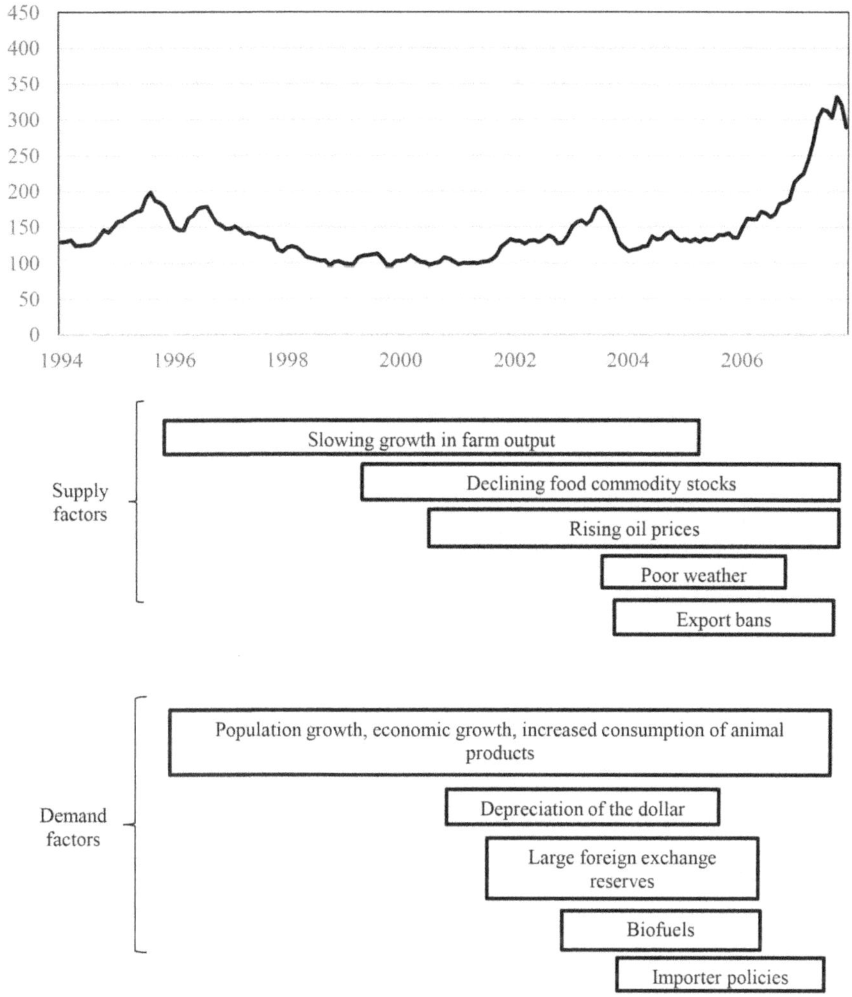

Fig. 3.6. The factors behind the food price spike of 2008 Average of four crops (maize, rice, soyabean and wheat). (Trostle (2011); World Bank commodity price data (the pink sheet)). January 2002 = 100

for farm animals, while soyabeans are a very important source of feed for farm livestock. The department examined the various forces that influence the average price of these four grains. The price cycle is shown in Fig. 3.6, together with the long-term and short-term factors behind it.

3.5.1 Long-term causes on the supply side

After their peak in 1996, food commodity prices declined as fast as they had risen and settled at a low level around the year 2000. They remained at this level for several years. From 2002, they gradually began to rise. Let us consider what was happening on the supply side.

The first factor was agricultural output – the physical quantity (i.e. tonnes) of commodities produced by farmers. During the cycle, world agricultural output grew at an average rate of about 1.3% per year. However, this rate of growth of agricultural output was less than before. Between 1970 and 1990, world agricultural production had been rising at an average 2.2% per annum. In other words, for 20 years from 1970, world agricultural production had been rising at 2.2% per annum but then slowed to 1.3%. Why did this slowdown happen?

The quantity of a crop that a farmer produces depends on two factors: the area of land that he decides to plant (hectares, acres) and the yield of the crop (tonnes/hectare, bushels/acre). Yield is determined by the farmer's methods (e.g. how much seed is planted, how much fertilizer is applied) and by the benevolence, or otherwise, of nature (e.g. too much rainfall or just sufficient or too little).

When one takes a long-term historical view of farming, one finds that the huge increases in food production have come from an expansion of the area farmed rather than from an increase in crop yields. Since the dawn of agriculture, humanity has gradually converted forest, woodland and grassland to farmland.

In his book for the Earth Policy Institute at Rutgers University, Lester Brown (2012) notes that:

> From the beginning of agriculture until the mid-twentieth century, growth in the world grain harvest came almost entirely from expanding the cultivated area...It is only within the last 60 years or so that rising yields have replaced area expansion as the principal source of growth in world grain production....Since 1950, over 93% of world grain harvest growth has come from raising yields. Expanding area accounts for the remaining 7%.

However, around the mid-1990s, in some important farming countries, this growth in yield appears to have been almost entirely arrested, as shown in Figures 3.7–3.9. This is the case, for instance, in France, Germany and the USA (wheat) and in China and Japan (rice). The figures show the plateauing of the yields of these crops in these countries.

Figure 3.7 shows how wheat yields rose rapidly in France and Germany from the 1960s. But beginning around the mid-1990s, this growth appears to have been almost entirely arrested. Although not shown, the same trend has taken place in other European countries, including the UK.

Figure 3.8 shows the situation regarding the most important variety of wheat in the USA (hard red wheat). Yields have been more or less stagnant since the mid-1990s, although the high yield in 2016/17, of 49.5 bushels/acre, may indicate the recovery of yield growth.

As a third example, Fig. 3.9 shows how rice yields in Japan were growing until the mid-1990s but have since reached a plateau. The same figure

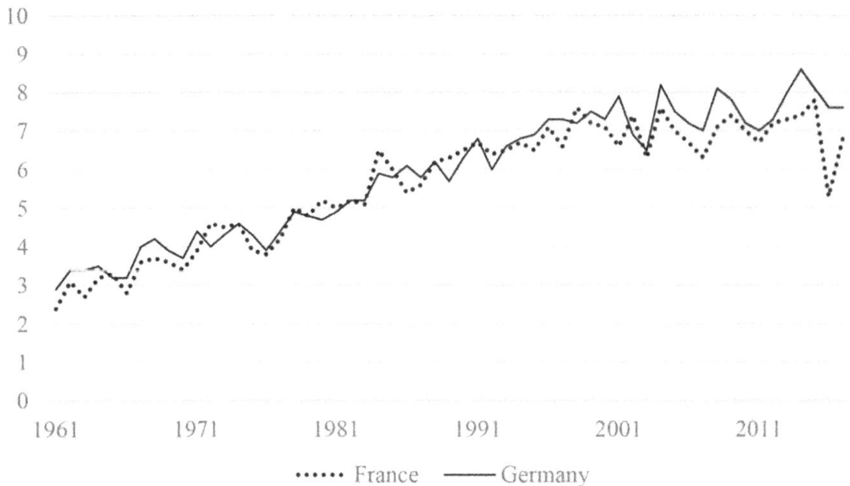

Fig. 3.7. Wheat yields in France and Germany (tonnes/ha). Source: FAOSTAT.

shows the rice yields in China, which are still growing but at a reduced pace compared to the 1960s, 1970s and 1980s.

This arrest in the growth rate of crop yields is a matter of considerable concern. Since the dawn of farming, Man has produced more food with every passing year. Up to the 1950s, this was mainly because he brought more land under the plough. Since the 1950s, it is because he has increased the yields of his crops. Many scientists and economists assumed that crop yields would continue to grow at a rate of around 2% per annum into the future, as they had done since World War II. But since the mid-1990s, in some countries at least, it seems that crop yields are no longer increasing. They have stagnated.[7]

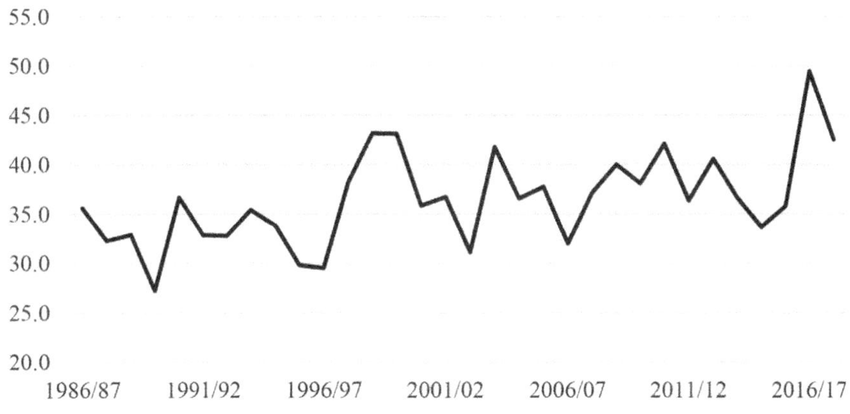

Fig. 3.8. US wheat yields of hard red winter in bushels/acre. (*Wheat Yearbook* database, USDA)

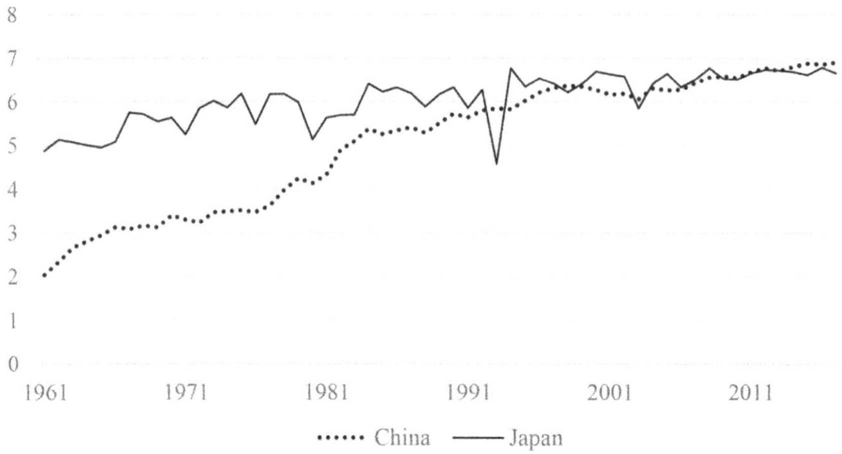

Fig. 3.9. Rice yields in China and Japan (tonnes/ha). (FAOSTAT)

The slowdown in agricultural output may well be explained by the stagnation in crops yields, illustrated in Figures 3.7– 3.9. But of relevance to the price of grains is the fact that the rate of growth of agricultural output fell below the rate of growth of the world's population. The number of stomachs in the world was increasing faster than the number of bags of food. The rate of increase in agricultural output during the 1996–2008 cycle was 1.3% per annum while the rate of increase of population was higher at 1.8% per annum. It meant that food commodity markets were gradually tightening; food was becoming scarcer (Trostle, 2008a, p. 5).

The second factor on the supply side was the level of stocks of grain. The price cycle started with a low level of grain stocks in many countries, brought about by 20 years of abundant world supplies, low grain prices and a perhaps naïve belief that the free market could supply the world's food needs, obviating the need for governments to hold grain reserves. But world stocks declined further during the cycle. The stocks-to-use ratio for aggregate global grains and oilseeds fell to less than 15% in 2008, the lowest level since 1970 (Trostle, 2008b). This was less than the 20% level that many observers consider to be the minimum safe level if the market is to function efficiently and if participants are to have confidence that supplies of grain will be forthcoming on the world market.

Regarding the fall in wheat stocks, this was precipitated because farmers switched some land from wheat to grow oilseeds. This switch from wheat to oilseeds took place mainly in Argentina, Canada, the European Union, Russia and Ukraine. The eight largest wheat exporters expanded their area under rapeseed and sunflower by 36%, amounting to 8 million hectares between 2001 and 2007, while the area under wheat fell by 1%. This reduced the global wheat harvest by some 26 million tonnes per year. The cumulative wheat production potential of the land taken out of wheat was 92 million tonnes over the period 2002 to 2007. To fill the gap in wheat supply, wheat stocks were thus drawn down (Mitchell, 2008, p. 11).

The third factor acting on the supply side was rising oil prices. Crude oil prices rose sharply from early 2002 to 2008. Much of the rise in price reflected increased demand caused by robust world economic growth and rapid manufacturing growth in China, India and other Asian countries. Higher fuel costs affect the price of food commodities in two ways. First, farmers have to pay more for their diesel, fertilizers and pesticides, these all being derived from oil. This pushes up farmers' production costs. The increased costs of ploughing, seeding, weeding and harvesting are passed down the marketing chain in the form of higher prices to the consumer.

Secondly, importers of food commodities have to pay higher ocean freight rates. As food commodity prices reached their peak in 2008, dry bulk freight rates increased even more dramatically, rising more than 350% between 2006 and 2007. High oil prices, increased demand for ocean shipping and slow growth in the availability of dry-bulk shipping vessels all contributed to higher freight rates.

Slow agricultural growth, a drawdown of stocks and high energy prices were long-term factors that caused supply to contract.

3.5.2 Long-term causes on the demand side

While supply was steadily contracting (falling), demand was gradually expanding. Five factors were acting on the demand side.

First, the world's population was growing. In 2002 the population of the world had just turned 6 billion and 75 million more mouths were being added each year. The majority of these extra mouths were in Africa, the Middle East and Asia. The growth in the world's population meant that there was greater demand for food.

The second factor was increased incomes and the consumption of animal products. Not only was the number of people in the world increasing, but so were people's incomes. This was particularly the case in Asia. With more money to spend, Asians began to consume more animal products: eggs, milk, cheeses, yogurt and meats. These are obtained from cows, sheep, pigs and poultry, often fed on maize and soyabeans. It requires much more grain to provide a given food value if it is consumed via animals than if the grain is consumed directly. The rising demand for animal products and animal feed was the second factor that increased the demand for food.

Thirdly, at the same time, the US dollar was depreciating in value against other currencies. Between 2002 and 2008, it depreciated by about 35% against the euro (Mitchell, 2008, p. 15). When traded from one country to another, food commodities are customarily priced in US dollars. If the value of the dollar falls, the price to the buyer falls with it. This is because the buyer pays in his own national currency, which now buys more dollars than previously because the dollar is worth less. Thus a falling dollar increases the demand for food commodities by buyers (governments and private traders) outside the USA. This was a third long-term background condition that caused the gradual increase in prices of food commodities after 2002.

The fourth factor on the demand side was large foreign exchange reserves held by some countries. It allowed them to bid for greater supplies of food commodities than normal. By late summer 2007, some importers were aggressively contracting for imports of grains and oilseeds. Even though prices were at record highs, these importing countries were buying larger volumes, not less. Some countries that usually imported sufficient quantities of grain to meet their needs for the following three to four months began to contract for imports to meet their needs for the following five to ten months. The large foreign exchange reserves held by these importing countries enabled them to contract for their import needs regardless of how high the world price rose.[8]

The fifth factor was an increased demand for biofuels. Some countries had been producing biofuels for many years such as Brazil which had been producing bioethanol from sugar. But in other countries – especially in the USA and Europe – the large-scale production of biofuels was entirely new. It was a novel source of demand for food commodities.

In the early 2000s, that is during the middle of the price cycle, the USA and the EU passed legislation requiring oil companies to blend biofuels (bioethanol from maize in the US and biodiesel from oilseeds in the EU) with gasoline.[9] The motive for the USA was to reduce the country's dependence on oil imports from foreign countries. The motive for the EU was to reduce emissions of CO_2, mitigating damage to the atmosphere and slowing, at least to some extent, global warming and climate change.

As a result of this legislation, the production of biofuels increased greatly. In the EU, production of biodiesel was 0.28 billion gallons in 2001; by 2007 it was 1.78 billion gallons. This was a six-fold increase in as many years. Rapeseed was the primary feedstock followed by soybean oil and sunflower oil.

Production also increased dramatically in the USA. The legislation of 2005 mandated that 7.5 billion gallons be blended with fossil fuels. This was later raised to 15 billion gallons. If these mandates were to be filled by domestic production, ethanol production would have had to double and biodiesel production would have had to more than treble. At the world level, biofuel production increased at a heady rate of over 30% per annum from 2006 to 2008 (Trostle, 2011, p. 15).

Globally, the 2002–2008 increase in biofuel production – ethanol in the USA and Brazil and biodiesel in the EU, Argentina and Brazil – played a role in raising prices for the feedstock grains: maize, sugar, rapeseed and soybeans, as well as for other crops.

These were the long-term factors – on both the supply side and on the demand side – causing grain prices to gradually increase from their low point in 2002. Supply was expanding but not as fast as demand. The market was gradually tightening. The conditions were ripe for a short-term shock or an anomaly in the market to cause a sudden increase in prices.

3.5.3 Short-term market shocks and anomalies

On the supply side, there was bad weather during 2006 and 2007. Droughts, floods, high temperatures and freezes affected agricultural yields in many countries. In 2007 alone, northern Europe suffered floods at harvest, while droughts enveloped south-eastern Europe, Ukraine, Russia, Turkey and Australia. Bad weather caused a drop in global average yields for grains and oilseeds. The lower harvests caused a drawdown of commodity stocks and further tightened the market.

Faced with rising grain prices, a number of countries decided to restrict or ban exports of grain. The intention was to ensure that grain stayed inside the country and to prevent any increase in the price of food for the citizens of the country itself. For instance, India banned exports of rice. This was to keep rice cheap in an election year. Indeed, it was in the rice market that we see the influence of government export restrictions most vividly. The debacle has been described by Tom Slayton of the Center for Global Development in Washington. He recounts the events as follows (Slayton, 2009, p. 1). In 2008,

> The world rice market was set aflame last spring and for several months it looked as if the trading edifice that had exhibited such resilience over the last two decades was going to burn to the ground.
>
> World prices trebled within less than four months and reached a 30 year inflation-adjusted high. Many market observers thought the previous record set in 1974 would soon be toast. The fire was man-made, not the result of natural developments. While the governments of India, Vietnam and the Philippines did not seek to set the world market on fire, that was the unintended result of their actions which threatened both innocent bystanders (low-income rice importers as far away as Africa and Latin America) and, ultimately, poor rice-consumers at home.
>
> Fortuitously, when the flames were raging at peak intensity, rain clouds appeared, the winds (market psychology) shifted and conditions on the ground improved, allowing the fire to die down.

By restricting exports of rice onto the world market, governments had caused a contraction of supply of rice onto the world market. This is an example of how, as grain prices rise, governments tend look after their own domestic interests and take measures to keep food prices low for their own consumers and citizens. Such measures can be very harmful for food-deficit countries which rely on imports.[10]

The final factor identified in Fig. 3.6 is importer policies. Some importers became aggressive buyers of grain. Trostle (2011, p. 21) has observed that as available global crop supplies declined and prices began to rise more quickly, some importing countries became concerned about their ability to import grain and meet the food needs of their population. Some of them began to aggressively make contracts with international grain traders for additional imports.

This ends our analysis of the rise in prices during the 1996–2008 cycle. It reveals that there was no single cause of the sudden increase in prices. A variety of factors were at play, some long-term, others short-term. Some factors caused demand to expand, other factors caused supply to contract.

3.6 The Role of Commodity Speculation

Whenever there is a sudden rise in the price of food, the finger of blame is often pointed at those who speculate on the agricultural futures markets. Such speculators buy and sell contracts on the futures exchanges of Chicago, London, Paris and elsewhere. The question is: does speculation cause food prices to increase? This is a vexed question that has been posed again and again over the decades. In the 1990s, the consensus among academics was that there was no clear evidence that the future markets had an adverse effect on the price of grain. For instance, in the 1990 edition of the standard reference book on agricultural prices, Tomek and Robinson (1990) stated that a number of empirical studies had attempted to measure whether the short-run volatility of cash prices had been influenced by futures trading. The studies found that either volatility was smaller or that it was unchanged after the introduction of futures trading. The authors quoted a study which showed that the daily range of futures prices for grain narrowed as the volume of speculation increased. In other words, the more speculation on futures contracts, the more stable was the price. They concluded that while futures trading may introduce brief episodes of aberrant price behaviour, the evidence suggested that the overall effect had been to reduce price volatility, or at least not to increase it.

Since the 1990s, the futures markets have evolved with, for instance, the introduction of high-frequency trading using computer algorithms and the growing participation of investment banks, hedge funds, pension funds and providers of commodity index products. What is the consensus of the academic profession now? Indeed, is there a consensus?

In the recent fifth edition of their classic book, Tomek and Kaiser (2014) observe that spot and futures prices have become increasingly variable since 2005. However, the role of futures markets, if any, in volatility remains difficult to appraise. Increased price volatility implies an increased demand for hedging and increased hedging and price volatility attract speculation. So speculative activity could be viewed as a consequence, rather than a cause, of price volatility. They conclude that empirical analyses of the direction of causation (that is, does price volatility cause speculation or does speculation cause price volatility?) can be difficult to interpret.

Speculation in futures contracts adds liquidity to the futures markets meaning that it is easier for hedgers to find a counter-party to a contract. This is a benefit that speculators bring to futures markets. Certainly, not all speculators make money – there is a view that most of them lose money. But on the question of whether speculation is a cause of price volatility, the jury is still out. There are many different views on the subject – and it is a complex area of research. It seems fairly safe to say that, if speculation is indeed a cause of price volatility (which is not certain), when its possible effects are set against the impact of the weather, of the level of stocks and of government policy, it is a relatively small and transient cause.

3.7 Conditions that Are Associated with Increases in Grain Prices

Among this thicket of factors that influence the price of grain – ranging from population growth, the consumption of biofuels, bad weather, the value of the dollar and so on and so forth – can we discern any general explanations of why prices of grain have been prone to sudden increases? Focusing only on wheat, there have been four pronounced increases in wheat prices since the abandonment of the International Wheat Agreement in 1971. These were in the mid-1970s, the mid-1990s, in 2007/8 and in 2011. A reading of some of the commentary and analyses of these periods suggests that in each of these four instances the price increase was associated with the following three factors: (i) bad weather in major grain-producing countries; (ii) a low level of grain stocks[11]; and (iii) trade measures (being import subsidies, export taxes and bans on the export of grain – all of which tend to increase the world price, as explained in Section 8.8).

Table 3.1 presents the evidence for this argument while Fig. 3.10 shows that when world wheat stocks are low, there is a tendency for the price of wheat to rise.

3.8 Have Grain Prices Arrived at a Turning Point?

3.8.1 The influence of global warming

There is a consensus that the world's climate is changing due to higher levels of CO_2 and other greenhouse gases in the atmosphere. What might this do to food production? While predictions about the effects of climate change are extremely difficult to make with certainty, the UN's Intergovernmental Panel on Climate Change (United Nations, 2014) has expressed its view that in Africa, crop production will fall due to heat and drought stress, that in Europe there will be increased restrictions on the use of water for

Table 3.1. The conditions that are associated with increases in grain prices.

	Bad weather	Low stocks	Government taxes and subsidies, export bans
Mid- 1970s	Winterkill and drought caused crop failures in 1971 and 1972 in the USSR, at that time the world's largest wheat producer.[12]	In the early 1970s, the stocks-to-use ratio was less than 20% and fell at one point to 15%.[13]	Japan, the EU, India, the USSR and China used policy to stimulate import demand. Canada, Australia and Argentina explicitly limited grain exports.[14]
Mid- 1990s	In the USA cereal crops declined by over 20% due to bad weather.[15]	In 1996, US wheat stocks were at their lowest level in 20 years. Worldwide supply of grain was at its lowest level for 35 years.[16]	The EU, which had been supplying 30% of world exports,[17] turned from subsidizing wheat exports to taxing them to ensure availability of wheat for animals. This reduced the supply of exports onto the world market.
2007/8	In 2006 and 2007: Australia, Ukraine and Russia drought for 2 years, Europe dry spring and harvest floods, USA late spring freeze, Canada hot and dry.[18]	Stocks-to-use ratio fell continuously from 30% in 1998 to 15% in 2008.[19]	Export bans, reduced import tariffs and food subsidies.[20] The supply of exports was reduced and, at the same time, the demand for imports was increased.
2011	Beginning in June 2010, weather around the world was either too hot, too cold, too dry or too wet. A heatwave in Russia destroyed crops on one third of farmland.[21,22]	Global stocks of grain declined.[23]	Various countries established export restrictions, export bans, reduced import tariffs and increased consumer subsidies to help offset rising food costs.[24]

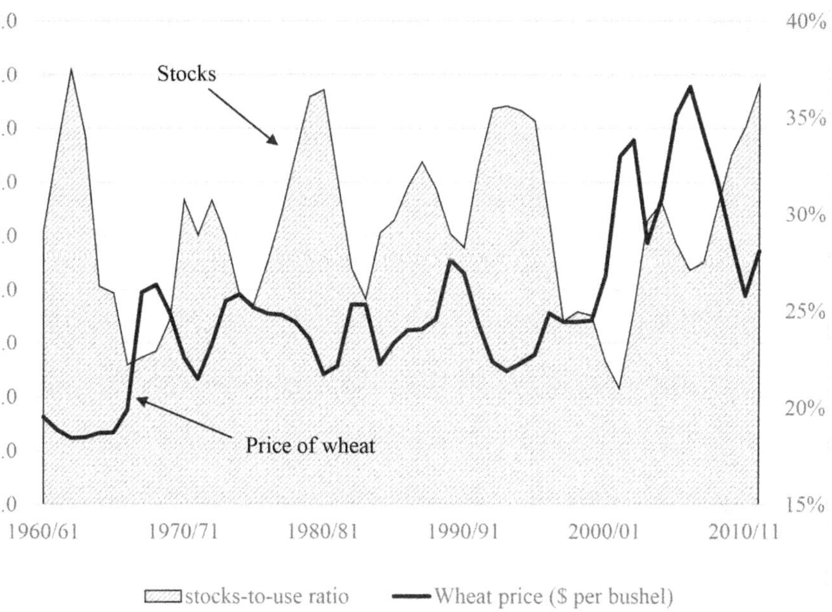

Fig. 3.10. When wheat stocks fall, the price increases. Wheat price on left-hand axis in dollars/bushel, stocks-to-use ratio on right-hand axis. (*Wheat Yearbook* database, USDA)

irrigation of crops, that in Asia there will be a higher risk of drought-related water and food shortages causing malnutrition, and that in South America there will be lower production of food. If these predictions are correct, there will be less grain produced in many countries of the world.

3.8.2 The longer-term supply-and-demand equation

Christopher Ritson has taken a very broad look at the prices of agricultural and food commodities over the past 60 years (Ritson, 2016). He has advanced the argument that up until recently the supply of grain has outpaced the demand for grain. New agricultural technology in the form of better seeds, the greater use of fertilizer and pesticides, more irrigation and so forth, have driven an expansion of supply. Demand for grain has, of course, expanded, too, driven by growth in population and growth in incomes. But supply has tended to expand at a faster rate than demand. Hence the price of grain has been forced down to a floor, punctuated by occasional and unexpected increases caused by the short-term disturbances and events such as we have described above.

Ritson has termed the period 1975–2005 – a period of 30 years during which grain prices were low – as the 'era of cheap food'. Grain prices were low and food was cheap because of one overriding factor: although both the supply of and the demand for grain were expanding, supply was expanding faster than demand. This meant that there was constant downward pressure on prices. It also meant that a poor harvest would have a distinct effect on the level of prices – there would be a rapid increase followed by an equally rapid fall, back to the normal level. Recently, however, it seems that something may have changed because the patterns of the past are no longer evident. The clue lies in the period from 2007. That year saw a sudden price increase but it was not followed by an equally sudden price decrease, as had occurred in the mid-1970s and at other times. Rather, prices remained high. Since 2007, grain prices have not fallen back to a floor but have generally stayed at a high level. This is new.

The question is: What might explain this persistence of high prices? We need to distinguish between short-term shocks and disturbances (such as poor harvests) and longer-term trends (such as farm technology, consumers' incomes and tastes, and so forth). It has generally been thought that short-term shocks reflect themselves in sudden increases and decreases in prices – in other words, as short-term volatility of prices (over weeks and months). Likewise, it has generally been thought that changes in long-term factors reflect themselves in smooth and gradual changes in prices. This has been the conventional wisdom in economic circles: short-term factors are associated with short-term price volatility while long-term factors play themselves out over years and decades.

This thinking may be – at least in part – mistaken. Could it not be the case that sudden increases in price are brought about, not only by short-term shocks but also by changes in the longer-term factors? Ritson argues that this is what may have happened since 2005. Looking just at the long-term factors, up until then, supply had been outpacing demand – hence the era of cheap food. But since 2005, it may well be that demand has been outpacing supply. In other words, the situation regarding the long-term factors has been reversed. A reversal in the long-term factors – by itself and without any influence of short-term factors – would be reflected not in a smooth and gradual increase in price but in a sudden and abrupt increase. A reversal of the long-term factors would have caused prices to increase suddenly – in the same way as a poor harvest causes a sudden increase in price. The question then is: Why did prices not fall back down? A poor harvest is a temporary event; if the following harvest is normal, prices fall back down. But a reversal in the long-term balance between supply and demand is not a temporary event; it is a more permanent event – hence prices have not fallen back down.

This switch may have been brought about by three factors underlying supply and demand since the early 2000s. One factor has affected supply and two factors have affected demand. On the supply side, as explained above, technology is now having less effect on crop yields (these were increasing rapidly in many countries until around 2000 after which they began to plateau). In other words, the supply may not expand as fast in the future as in the past. On the demand side, first, the world's population is still increasing, and secondly, incomes are increasing in Asia causing an expansion of demand for foods from grain-fed livestock. These two factors are driving forward demand. Demand for grain may expand faster in the future than in previous years.

This suggests that, henceforth, demand for grain will expand faster than supply. The consequence is that prices will tend to hit a ceiling. They may bounce along this ceiling with occasional falls due to short-term factors such as economic recessions and exceptionally good harvests. Whereas in the past sudden price increases resembled stalagmites ascending from a floor of a cave, in the future it may be that prices hit a ceiling from which they occasionally descend like stalactites. This is the compelling image drawn by Ritson. It suggests a very different future for grain prices, a future characterized more by shortages than by surpluses. The era of plentiful supplies may lie in the past, not in the future.

3.9 Conclusion

This chapter has shown how the prices of food commodities – and more particularly the prices of grain – have fluctuated since the 1960s. It has endeavoured to explain the main factors behind the sudden and unexpected increase in the price of grain that occurred in 2008. Finally, it has suggested that the world may be entering an entirely new era in which grain prices, rather than bouncing up from a floor, instead bounce down from a ceiling. But does it matter that grain prices are unstable? The following chapter examines the economic causes of volatile prices.

Notes

[1] The price series has been deflated (for an explanation of the index used, see FAO (2013)). This means that any increases in the market price of food that have been brought about by a fall in the value of money (inflation) have been stripped out. By stripping out inflation, the figure shows the real economic resources that consumers have had to exchange in order to acquire a standard basket of foodstuffs since 1961. For instance, we see that there is a sudden increase in price in the early 1970s. Because the price series has been deflated, the figure tells us that, during this period, consumers have had to give up a greater quantity of their real economic resources to acquire the same amount of food. Other things being equal, consumers were therefore worse off. If the price series had not been deflated – if it expressed nominal prices – the increase in the early 1970s could simply have been due to a fall in the value of money (i.e. due to a period of inflation). In this case, consumers' wages may have been increasing at the same rate as the price of food. Other things being equal, consumers would not have been worse off.

What does Fig. 3.1 tell us? The overall picture is that, in terms of the real economic resources that they have to expend in order to acquire food and assuming that all other things have remained the same, consumers are more or less in the same position now as they were 50 years ago. Of course, it is not the case that all other things have remained the same over the last 50 years. In particular, in most countries of the world, there has been a great deal of economic growth. Economic growth means that there are more economic resources available. Many consumers therefore have more economic resources now than they had 50 years ago. For these consumers, food is now cheaper relative to their incomes. They need to expend a smaller proportion of their real income to acquire the same basket of food. But for those consumers whose incomes have not increased, the real cost of food is the same as it was 50 years ago. The amount of real economic resources that these consumers need to expend in order to acquire their food is the same today as it was 50 years ago.

[2] It may be that food prices do not rise in real terms even if, in nominal terms, they do rise. For instance, wages may rise in line with the rise in the price of food. But this does not alleviate the plight of families faced with rises in the nominal price of essential foodstuffs. First, it can be that a rise in wages takes place several months after an increase in food prices. When measured over the course of a year, real food prices would not have increased. But the reality is that consumers have to buy food on a continuous basis during the year. It is of little consolation to consumers facing very high food prices that their wages may increase in several months' time. Secondly, food purchases cannot be delayed as can the purchases of many other items such as furniture, houses, cars and so forth. These can be purchased after a wage increase – when they are more affordable.

[3] I have used World Bank data, which go back to 1960, rather than IMF data, which go back only to 1980. The database is the Global Economic Monitor Commodities. The commodities are: Maize (US), no. 2, yellow, f.o.b. US Gulf ports, source: US Department of Agriculture; Rice (Thailand), 5% broken, white rice (WR), milled, indicative price based on weekly surveys of export transactions, government standard, f.o.b. Bangkok, source: US Department of Agriculture; Wheat (US), no. 1, hard red winter, ordinary protein, export price delivered at the US Gulf port for prompt or 30 days shipment, source: Bloomberg, USDA. The database gives average prices for each year. These I have removed, leaving only monthly prices.

[4] Compared to wheat and maize, relatively little rice is traded internationally. Most rice is consumed in the country in which it is grown. The figure refers only to that portion of the world's rice harvest that enters international trade.

[5] The late Frank Golay of the University of Chicago, described the International Wheat Agreement thus: In substance, it is a contractual obligation between five wheat exporting countries and 36 wheat importing countries, in which: 1. Each member exporting country, if called upon by the Wheat Council to do so, agrees to export a stated amount of wheat to member importing countries at the maximum price defined in the agreement. 2. Each member importing country, if called upon by the Wheat Council to do so, agrees to buy a stated amount of wheat from member exporting countries at the minimum prices defined in the agreement (Golay, 1950, p. 443).

[6] I have drawn principally on the work and writings of the following specialists, many of whom have decades of knowledge of agricultural markets: Philip Abbott

of Purdue University in Indiana, USA; Dr Derek Headey and Sheggen Fan of the International Food Policy Research Institute, Washington; Peter Timmer of Harvard University; Dr David Dawe of the Asian Office of the Food and Agriculture Organization; Dr Donald Mitchell of the World Bank, Washington, and Dr Ronald Trostle of the USDA, Washington. For rice, I have drawn upon Dr Tom Slayton, former publisher of The Rice Trader and the Rice Trade Report.

[7] See Ritson (2016) for a further discussion of world food security and an explanation of the stagnation of crop yields.

[8] Trostle (2008a, p. 22) has written: There have been very large accumulations of foreign exchange reserves held by oil-exporting countries (OPEC and Russia) and by countries with large non-oil trade surpluses (China, Japan, and other Asian countries). Countries holding these large foreign exchange reserves are able to import large volumes of food commodities in order to meet their consumption needs and allay their domestic food price inflation. In essence, they can bid supplies away from other traditional importers that do not hold significant foreign exchange reserves.

[9] In the USA, the Energy Policy Act of 2005 required the use of 7.5 billion US gallons of renewable fuel by 2012, and the Energy Independence and Security Act of 2007 raised the standard to 36 billion gallons of annual renewable fuel use by 2022.

[10]This was not the first time that the actions of some governments, in times of global food scarcity, exacerbated the plight of other countries. It had already happened in the mid-1970s, described by Thomas Grennes of the University of North Carolina (Grennes, 1984, p. 350): 'The initial excess demand for grain raised world prices, but the major importing and exporting countries refused to permit domestic grain prices to respond to external conditions. Consequently, import demand was greater and export supply was smaller than they would have been in the absence of insulating policies. All the major importing countries, including Japan, the European Economic Community, India, the USSR and China used government policy to stimulate import demand. At the same time, Canada, Australia and Argentina explicitly limited grain exports. The United States was the only major grain exporter to refrain from using export trade restrictions. The result of these trade policies was to make grain inventories less adequate than they otherwise would have been. Because of the instability that was partly induced by trade policies in the rest of the world, prices were destabilised in the United States.'

[11]The role of stocks in the determination of world wheat prices is discussed by Carter, C. et al. in Antle, J. and Young, V. (1999, p. 100). They concluded that 'the 1960s to the mid-1980s was an era in which the governments of the major grain-exporting countries (the USA, Canada and the EU) held substantial stocks of grain. During this period, low stocks-to-use ratios tended to result in sharp increases in world grain prices. In the late 1980s and 1990s....it was still the case that decreases in stocks-to-use ratios generally resulted in price increases....'

[12]Trager, 1973, p. 7 and 114

[13]Trostle (2008c): graph number 9: Total world grain and oilseed stocks

[14]Grennes (1984), p. 350

[15]Food And Agriculture Organization of the United Nations (1996) The State Of Food And Agriculture, Section 6: International Agricultural Prices. Rome. Downloaded at: http://www.fao.org/3/w1358e/w1358e00.htm

[16]Carter, A. (1999). In Antle, J. and Smith, V., p. 79
[17]Antle and Smith (1999). In Antle, J. and Young,V., p. 46
[18]Trostle (2008c): graph number 25: Adverse weather reduced crop production in 2006 and 2007
[19]Trostle (2008c): graph number 9: Total world grain and oilseed stocks
[20]Trostle (2008c): graph 27: defensive measures by exporters, defensive measures by importers
[21]The Guardian newspaper (UK) of 5 August 2010 reported as follows: 'Russia's prime minister, Vladimir Putin, has announced a ban on grain exports after millions of hectares of crops perished in the worst drought in more than a century. High temperatures, lack of rain and wildfires have devastated more than a third of cultivable land in Russia, the world's fourth largest grain exporter. News of the ban pushed wheat prices to a 23-month high on commodities markets and raised concerns about a boost in food prices worldwide.'
[22]Trostle (2011), p. 17
[23]Trostle (2011), p. 17
[24]Trostle (2011), p. 17

References

Antle, J.M. and Smith, V.H. (1999) *The Economics of World Wheat Markets.* CAB International, Wallingford, UK.

Brown, L. (2012) Full planet, empty plates: the new geopolitics of food security. Available at: www.earthpolicy.org

FAO (2013) FAO's food price index revisited. In Special Features of *Food Outlook*, November issue, 68–71.

Golay, F.H. (1950) The International Wheat Agreement of 1949. *The Quarterly Journal of Economics* 64(3), 442–463. DOI: 10.2307/1884559.

Grennes, T. (1984) *International Economics.* Prentice Hall, New Jersey.

Mitchell, D. (2008) *A Note on Rising Food Prices,* Policy research working paper 4682. Development Prospects Group, World Bank, Washington.

Ritson, C. (1977) *Agricultural Economics: Principles and Policy.* Crosby Lockwood Staples, UK.

Ritson, C. (2016) The concept of food security and the future of world food prices. Discussion paper 37. Centre for Rural Economy, Newcastle University, UK.

Sewell, T. (1992) *The World Grain Trade.* Woodhead-Faulkner, New York and London.

Slayton, T. (2009) Rice crisis forensics: how Asian governments carelessly set the world rice market on fire. Working paper 163. Centre for Global Development, Washington, DC.

Tomek, W.G. and Kaiser, H.M. (2014) *Agricultural Product Prices,* 5th edn. Cornell University Press, Ithaca, New York.

Tomek, W.G. and Robinson, K.L. (1990) *Agricultural Product Prices,* 3rd edn. Cornell University Press, Ithaca, New York.

Trager, J. (1973) *Amber Waves of Grain: The Secret Russia Wheat Sales that Send American Food Prices Soaring.* Arther Fields Books, Inc., New York.

Trostle, R. (2008a) *Global agricultural supply and demand: factors contributing to the recent Increase in food commodity prices,* Revised July 2008. WRS-0801 Economic Research Service, USDA, Washington, DC.

Trostle, R. (2008b) Fluctuating food commodity prices: a complex issue with no easy answers. *Amber Waves.* 1 November, 2008, United States Department of Agriculture, Washington, USA.

Trostle, R. (2008c) *Factors Contributing to Recent Increases in Food Commodity Prices, Printout of Graphs and Charts.* USDA Economists Group, 10 June 2008, Economic Research Services, USDA, Washington. Available at: https://www.ers.usda.gov/publications/pub-details/?pubid=40464

Trostle, R. (2011) *Why Have Food Commodity Prices Risen Again?* WRS 1103, Outlook edition of June 2011. United States Department of Agriculture, Washington.

United Nations (2014) *Working Group II Contribution to the Fifth Assessment Report of the Intergovernmental Panel on Climate Change,* Assessment Box SPM 2, Regional Key Risks. United Nations, New York.

4 The Root Cause of Unstable Grain Prices

4.1 Introduction

To solve a problem, one has to know its cause. In this case, the problem is the instability of the market price of grain. If we can discern why prices sometimes increase markedly with little warning and at other times fall like a stone, we shall be halfway towards a solution.

The discussion in the previous chapter endeavoured to explain the reasons for changes in price in terms of specific, real-world events and trends, such as falling crop yields, the increase in the income of the people of Asia, and so forth. In this chapter we do not talk of real-world events. Instead, we discuss the price of grain in terms of concepts. Abstracting from the practical to the conceptual helps to capture the important eventualities that may arise.

The purpose of this chapter is to explain, in conceptual terms, why the price of grain and other agricultural commodities is unstable. In order to do this, we deploy an economic model of the market. Since we wish this book to be accessible to readers of all disciplines – including those who have not studied economics – we begin our explanation with the standard market model of supply and demand. Some readers will already be familiar with this model. They can proceed straight to Section 4.4 'Modifying the conventional form of the market model', where we explain how the standard model needs to be modified.

The model of the market helps us structure our thinking, organize our information and explore various scenarios. Section 4.2 presents the basic elements of the conventional form of the model of the market. This is the form that applies to goods in general – shirts, mobile phones, cars and so forth. Most readers will be familiar with this form of model of the market. In Section 4.3 we ask what determines the extent to which the price of a good changes when the market settles at a new point of economic equilibrium.

In order to apply to grain, the conventional form of the market model needs to be modified (4.4). The following section (4.5) uses the model of the grain market to show that price volatility is caused by shifts in demand and supply and, moreover, that small shifts generate big fluctuations in price. In Section 4.6 we ask if it is possible to prevent shifts in demand and supply. Section 4.7 draws conclusions.

4.2 The Basic Elements of the Conventional Model of the Market

4.2.1 The demand for a general good

In economic terms, the 'demand' for a good is the quantity that consumers are able and willing to buy with money, during a given time period, over a range of different prices, all other things remaining constant.

The demand for a good is not the quantity that someone may desire to consume were the good offered for free and they did not have to pay for it. Neither is it the quantity that a person needs to consume, for instance to remain in good health (e.g. x kg of food per day). The demand for a good is its economic demand, in other words, demand backed by the expenditure of money.

It tends to be the case that the higher the price of the good, the less consumers buy. This is illustrated by a 'demand schedule' as shown in Table 4.1.

4.2.2 The supply of a general good

The supply of a good is the quantity that producers (in our case farmers) are able and willing to offer on the market, during a given time period, over a range of different prices, all other things remaining constant. It tends to be the case that, for most goods, the higher the price that producers receive, the more they will produce and offer on the market during a particular time period. This tendency can be demonstrated in the form of a 'supply schedule' as shown in Table 4.2.

Table 4.1. A demand schedule for a good.

Market price ($ per unit)	Quantity of good bought by consumers (number of units)
1	40
2	30
3	20
4	10

Table 4.2. A supply schedule for a good.

Market price ($ per unit)	Quantity of good supplied by producers (number of units)
1	10
2	20
3	30
4	40

4.2.3 The point of economic equilibrium

Supply and demand are presented in the conventional form of the model of the market, as in Fig. 4.1.

It shows the conventional market model for goods in general. We see that supply and demand balance each other at a point of economic equilibrium. This is defined by the equilibrium price and the equilibrium quantity. In this particular example, the equilibrium price is $2.5 per unit of the good. The equilibrium quantity is 25 units. It means that producers will receive $2.5 for each unit of the good that they produce and that they will produce 25 units. As for consumers, they will pay $2.5 for each unit of the good that they buy and they will buy 25 units.

At the point of economic equilibrium, the market 'clears', which is to say that there is neither a shortage nor a surplus of the good. Whatever is produced is purchased and the quantity that consumers wish to buy is

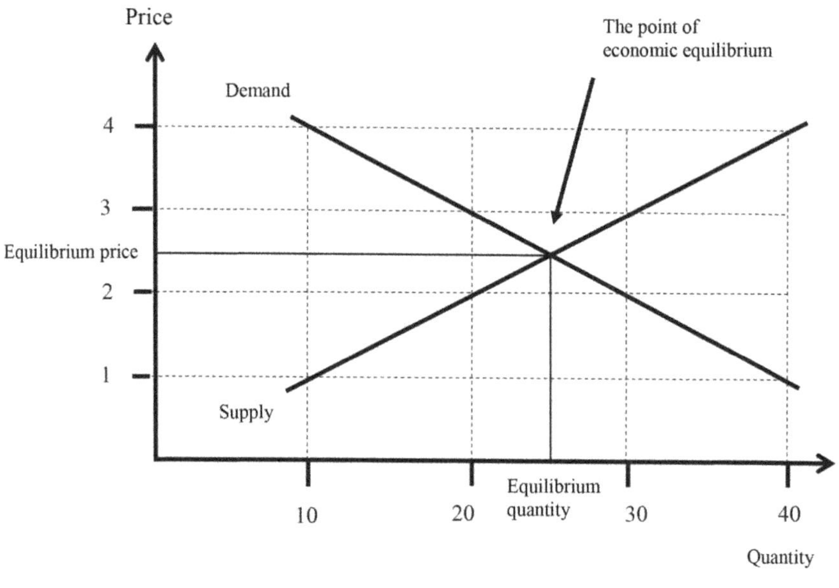

Fig. 4.1. The conventional form of the market model.

exactly matched by the quantity that producers are willing and able to produce. The market is in balance.

In some circumstances, the price at which a good is brought and sold (the market price) may be above, or below, the equilibrium price. In this case, there is no point of economic equilibrium and the market is not in balance. If the market price is above the equilibrium price, the quantity offered for sale will exceed the quantity that consumers are willing to buy. There will be a 'market surplus'. If the opposite situation pertains – that is, the market price is below the equilibrium price – then the quantity demanded by consumers will exceed the quantity that producers are willing to offer for sale. There will be a 'market shortage'.

If producers and consumers are free to alter their prices, then in a situation of market surplus, producers will lower their prices to dispose of unsold goods. In a situation of market shortage, consumers will bid up prices to obtain the quantity they wish to buy. In this way, market forces drive the market price to the level of the equilibrium price. A new point of economic equilibrium is established, the quantity supplied equals the quantity demanded and the balance of the market is restored.

4.2.4 Shifts in demand

As we have stated above, the demand for a good is the quantity that producers are willing and able to offer for sale at different market prices, during a specified time period, all other things remaining constant. In addition to how much a good costs in the shop (its market price), there are many other factors that influence the quantity of a good that consumers buy. Two of the most important are the level of consumers' incomes and their tastes and preferences. In the schedule above (Table 4.1) the level of consumers' incomes is assumed to be the same over the range of prices that we have considered. The same applies to the tastes and preferences of consumers. Their tastes are assumed to be the same whether the price of grain is $1 per bushel or $4 per bushel, or anything in between. This is what is meant by 'all other things remaining constant'.

In the real world, of course, 'all other things' do not necessarily remain constant; they change – people's incomes can increase, meaning that they have more money to spend on goods. Equally, incomes can decrease. Tastes and preferences can change due to all sorts of reasons: the prevailing conventions of society, the season of the year, and so on and so forth.

Such changes cause the demand curve to 'shift'. Let us suppose that there is rapid economic growth which brings about an increase in consumer income. The likely result is that consumers will demand more of the good. In this case, we say that there has been an *expansion* of demand. On the other hand, a recession will cause incomes to fall. Consumers will cut back on their consumption of the good. In that case we say that there has been a *contraction* of demand for the good.

Table 4.3. An expansion and a contraction of demand.

	Quantity of good demanded by consumers (number of units)		
Market price ($ per unit)	At original level of income	At increased level of income (an expansion of demand)	At decreased level of income (a contraction of demand)
1	40	50	30
2	30	40	20
3	20	30	10
4	10	20	0

Table 4.3 gives an example of the quantities that are demanded when there is an expansion of demand – due, for instance, to a higher level of consumer income – and when there is a contraction of demand – due, for instance, to a lower level of consumer income.

Figure 4.2 shows the new demand curves. Each and every time that there is a shift in demand, the market finds a new point of equilibrium. It finds a new balance between these two market forces of supply and demand. There is a new equilibrium price and a new equilibrium quantity.

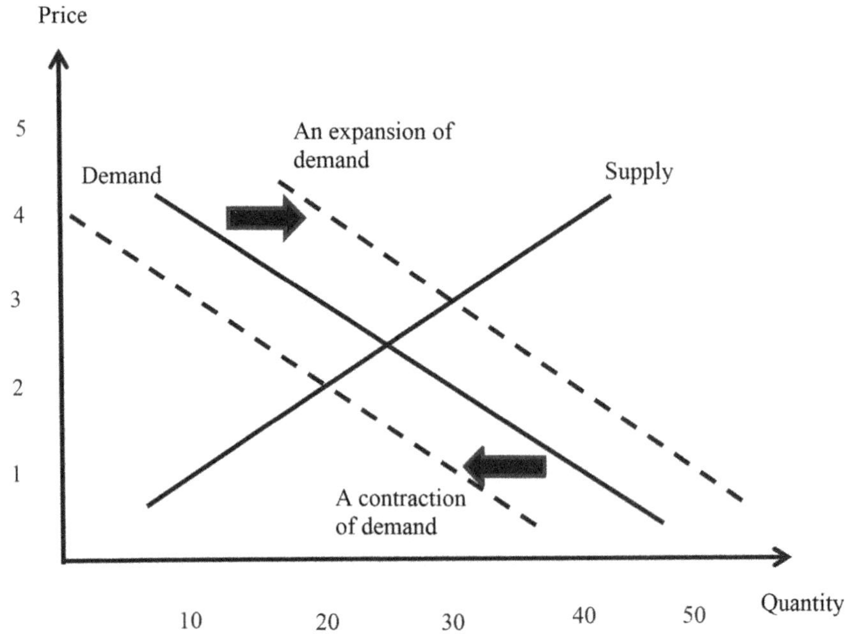

Fig. 4.2. Shifts in demand.

Table 4.4. An expansion and contraction of supply.

Market price ($ per unit)	Quantity of good supplied by producers (number of units) At original cost of labour	At decreased cost of labour (an expansion of supply)	At increased cost of labour (a contraction of supply)
1	10	20	0
2	20	30	10
3	30	40	20
4	40	50	30

4.2.5 Shifts in supply

Turning now to the supply of a general good, the same consideration applies – the schedule given in Table 4.2 above is the supply that producers are willing and able to offer to the market *all other things remaining constant.* It is only the market price that varies over the range of prices that we are considering. Other things of relevance to producers – the cost of raw materials, the cost of hiring labour, the taxes they have to pay on their profits, to take just a few examples – remain the same.

Of course, these factors do change with time and circumstances. When they do so, there is a shift in supply. We saw above that demand can expand and contract. Similarly, supply can expand and contract. Let us take, for instance, a change in the cost of labour to make a good. The cost of labour may increase or it may decrease. Table 4.4 shows an expansion and contraction of supply due to a change in the cost of labour. A decrease in the cost of labour leads to an expansion of supply. An increase in the cost of labour leads to a contraction of supply. Figure 4.3 shows the new supply curves.

Just as the market settles at a new equilibrium each time there is a shift in the demand curve, so the same thing happens whenever there is a shift in the supply curve. As can be seen from the figure below, when there is an expansion of supply, the new equilibrium is at a lower price. When there is a contraction of supply, the new equilibrium is at a higher price.

4.3 The Elasticity of Demand and Supply with Respect to Price

The two concepts of price elasticity of demand and price elasticity of supply lie at the heart of any economic analysis. Indeed, they are especially important in the context of volatility of food prices. They explain why food prices can suddenly escalate and suddenly fall. We now explain these concepts in the context of ordinary, general goods and apply them later to agricultural commodities and food.

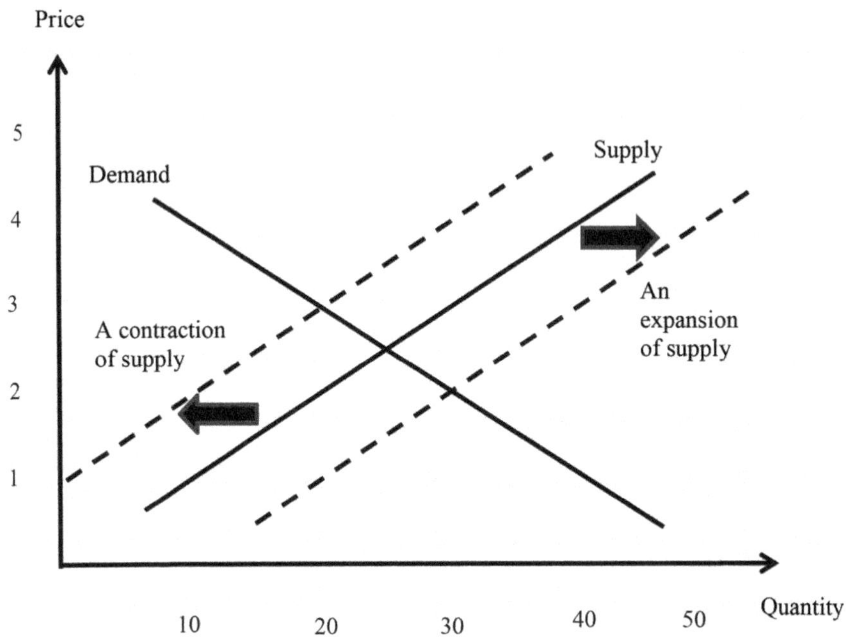

Fig. 4.3. Shifts in supply.

Elasticity is akin to flexibility. It is a measure of the extent to which quantity is flexible relative to the extent that price is flexible. Whenever there is a shift in either supply or demand, the market seeks a new point of equilibrium. Only two things can adjust – the quantity of the good that is brought and sold (the equilibrium quantity) and the price at which the good is bought and sold (the equilibrium price). Either or both of these two variables has to change for the market to find its new balance. The question is: Which one changes? Do they both change? Or does one stay where it is, while the other shows flexibility and moves to the full extent necessary for the market to find its new point of balance? Price elasticity is a measure of the relative flexibility of price and quantity. Let us first consider the concept of elasticity of demand with respect to price.

The elasticity of demand with respect to price is defined as the proportionate change in quantity demanded divided by the proportionate change in price. For instance, if price of a good increases by 20% and this induces a reduction of 10% in the quantity of the good that consumers demand, the elasticity of demand with respect to price is 10% divided by 20% which equals 0.5. Of course the price of a good may fall. Let us say the price of a second good falls by 50%. Let us also say that, as a result of the lower price, consumers increase the quantity that they demand by the same proportion, 50%. The elasticity of demand with respect to price of this second good is 50% divided by 50% which equals 1.

Figure 4.4a below shows a demand curve which is vertical. It describes a type of good for which a change in price does not evoke any change in the quantity that consumers demand. Consumers demand the same quantity of the good, irrespective of the price that they have to pay for it. In economic terminology, the demand for this good is 'perfectly inelastic' with respect to price. It means that when a shift in supply requires the market to find a new point of equilibrium, the market does so by adjusting the price alone. None of the adjustment is taken up by the quantity.

The expansion of supply from S1 to S2 causes the market to settle at a new equilibrium. The fact that the demand for this particular good is perfectly inelastic means that the entire economic adjustment is forced upon the price. The equilibrium price was P1. Following the expansion of supply, the new equilibrium price is P2, substantially less. The equilibrium quantity does not change – it remains the same at Q1.

In graph (b), the demand curve is steep. It describes a good for which the quantity demanded by consumers *does* change if the price changes (compare (a), where the quantity demanded did not change). The demand for this type of good is said to be 'inelastic' with respect to price, but not perfectly inelastic as in (a) above. This means two things. It means, first, that when a shift in supply requires the market to find a new point of equilibrium, the market does so by adjusting *both* the price and the quantity demanded. The price falls from P1 to P2, while the quantity rises from Q1 to Q2. Secondly, it means that, in proportionate terms, the price is adjusted more than the quantity is adjusted.[1]

In graph (c), the slope of the demand curve is shallow. It describes, as (b) does, a good for which the quantity demand changes if the price changes. The demand for this type of good is said to be 'elastic' with respect to price. Again, it means two things: first, that when a shift in supply requires the market to find a new point of equilibrium, the market does so by adjusting *both* the price and the quantity demanded. Secondly, it means that, in proportionate terms, the price is adjusted *less* than the quantity is adjusted.

In graph (d), the demand curve is horizontal. It describes a good which is demanded at only one price. If the price rises above this level, consumers do not demand any of the good. The demand for this type of good is said to be 'perfectly elastic' with respect to price. When the market for a good with a perfectly elastic demand seeks a new point of equilibrium, the adjustment is made entirely by the quantity demanded. In the figure, this has increased from Q1 to Q2. The price is not adjusted at all – it remains at P1.

In summary, whenever there is a shift in supply, the more demand is inelastic, the greater will be the change in price, as can be seen in Fig. 4.4. In (a), the demand is perfectly inelastic and the price fall is substantial. In (b), the demand is elastic and the price fall is considerable, but is less than in (a). In (c), the demand is elastic and the fall in price is rather modest. Lastly, in (d), the demand is perfectly elastic and there is no fall whatsoever in price. The implication, then, is that it is extremely helpful, in any

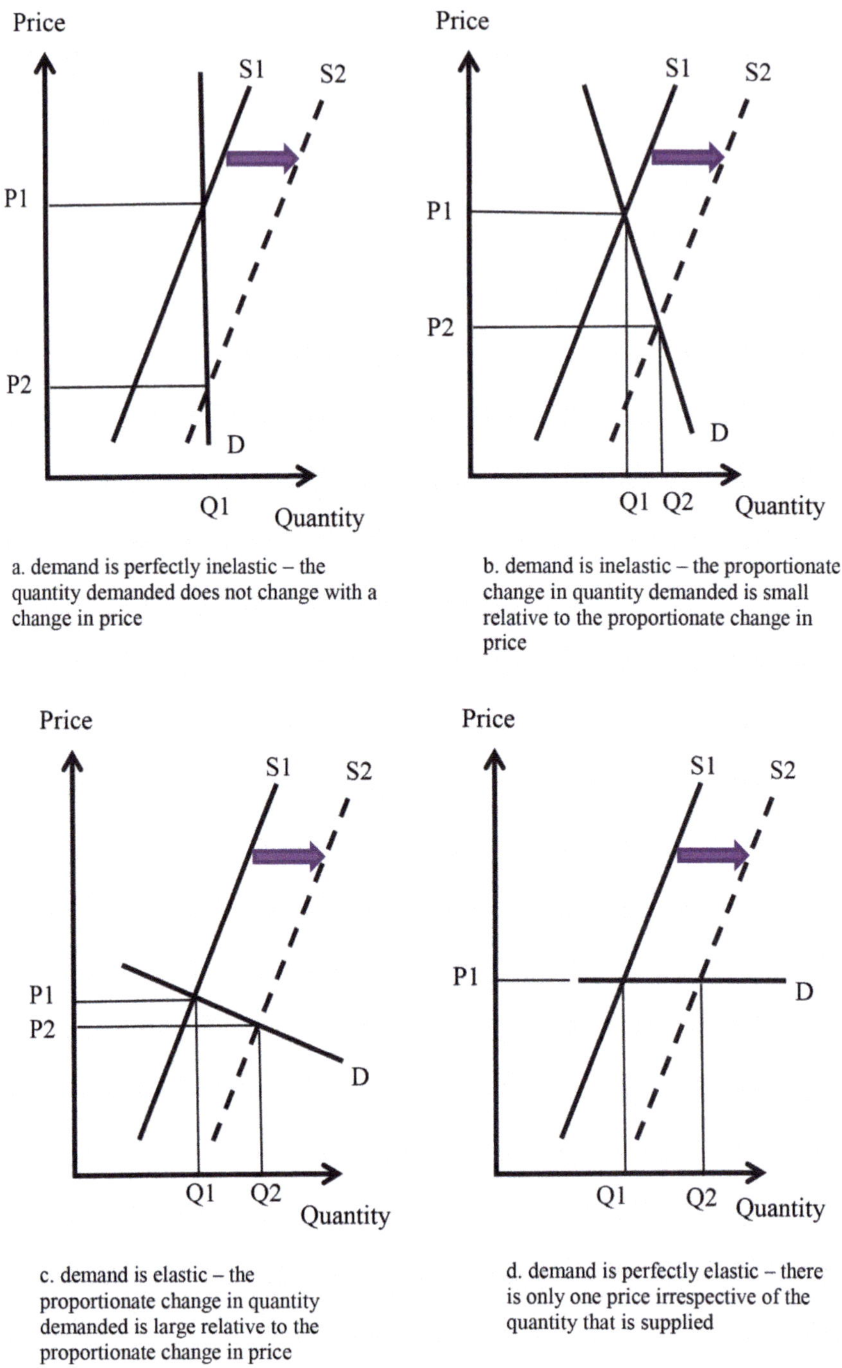

a. demand is perfectly inelastic – the quantity demanded does not change with a change in price

b. demand is inelastic – the proportionate change in quantity demanded is small relative to the proportionate change in price

c. demand is elastic – the proportionate change in quantity demanded is large relative to the proportionate change in price

d. demand is perfectly elastic – there is only one price irrespective of the quantity that is supplied

Fig. 4.4. The price elasticity of demand with respect to price.

discussion of prices and their stability, to know whether the demand for the good is elastic or inelastic with respect to price.

This explains the concept of elasticity of demand with respect to price. We need also to familiarize ourselves with the concept of elasticity of supply with respect to price. Assume that consumers have higher incomes and that they wish to purchase more of a good. The demand curve shifts to the right. The supply curve remains unchanged – it does not shift its position. The economy will seek a new point of equilibrium. Again, the only two things that can change are the equilibrium price and the equilibrium quantity. The elasticity of supply is a measure of the extent to which the price moves relative to the extent that the quantity moves.

The elasticity of supply with respect to price is defined as the proportionate change in quantity supplied divided by the proportionate change in price. For instance, let us say that an expansion of demand causes the equilibrium price to increase by 50%. Let us further say that this increase in price of 50% induces producers of the good to increase the quantity that they supply by 20%. In that case, the elasticity of supply with respect to price of the good is 0.4, this being 20% divided by 50%. Just as the demand for a good can, in principle, vary between perfectly inelastic and perfectly elastic with respect to price, so too can the supply of a good vary between perfectly inelastic and perfectly elastic with respect to price. Fig. 4.5 shows four possibilities.

In all four instances, there is an identical expansion of demand (the demand curve shifts the same distance to the right). This occasions a new point of economic equilibrium. In (a), the adjustment is entirely on the price, which rises from P1 to P2. None of the adjustment has been made on the quantity supplied. This is because, in this particular case, the quantity of the good that is supplied does not respond to a change in price. In economic terms, the supply is perfectly inelastic with respect to price. In (b), the supply of the good is inelastic, while in (c) it is elastic. Lastly, (d) shows a good for which the supply is perfectly elastic with respect to price.

Equipped with these two concepts, we are now in a position to explore how prices change when there are shifts in supply and demand. Figures 4.2 and 4.3 tells us the direction of a price change. Prices rise if there is an expansion of demand or a contraction of supply; prices fall if there is a contraction of demand or an expansion of supply. As for the extent of the price rise or price fall, this is determined by the price elasticity of demand (in the case of a shift in supply) and by the price elasticity of supply (in the case of a shift in demand). In other words, we have a model of the market for ordinary or general goods. This model enables us to structure our thinking as to what determines the price of a good and to explore various scenarios concerning supply and demand. This is the conventional model of the market with which many readers will be familiar. However, we cannot apply this model to grain and food, because they are not ordinary or general goods. They have some special

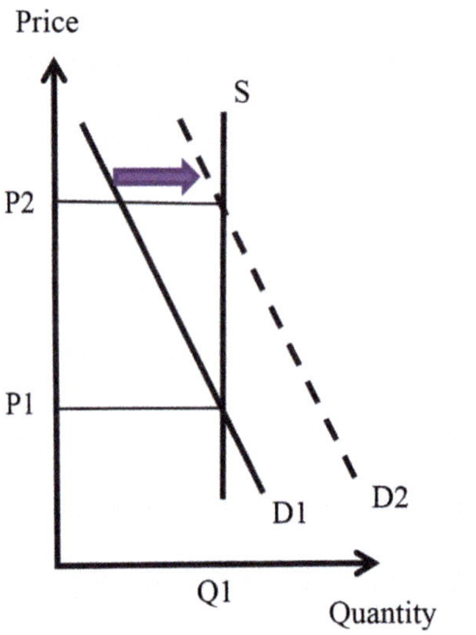

a. Supply is perfectly inelastic – the quantity supplied does not change with a change in price

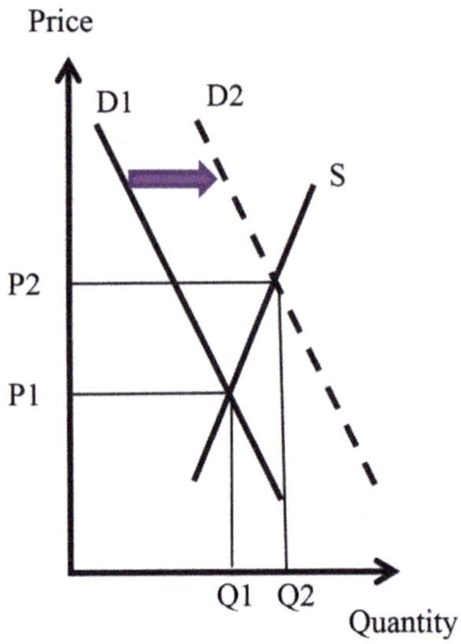

b. Supply is inelastic – the proportionate change in quantity supplied is small relative to the proportionate change in price

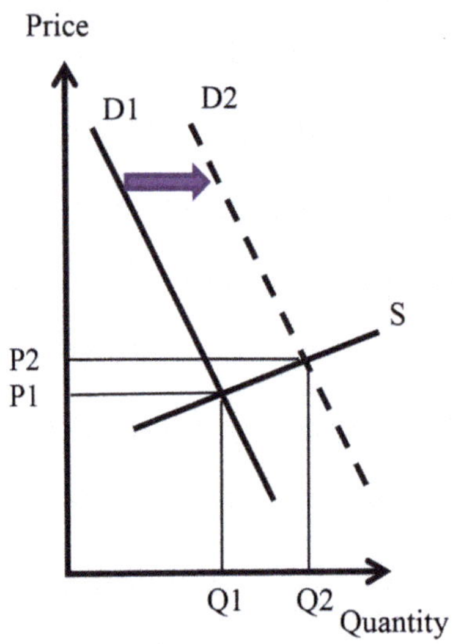

c. Supply is elastic – the proportionate change in quantity supplied is large relative to the proportionate change in price

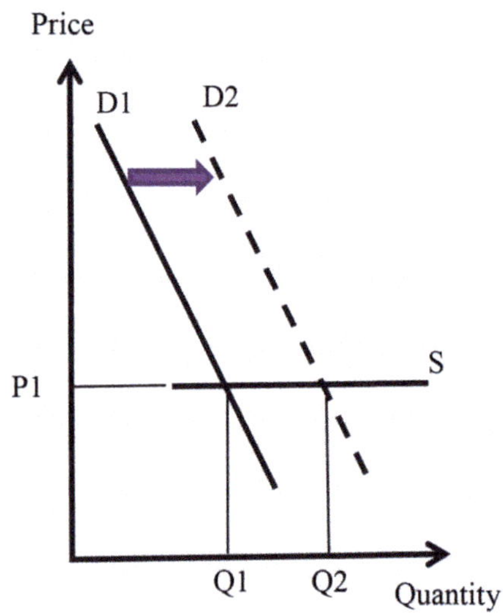

d. Supply is perfectly elastic – there is only one price irrespective of the quantity that is demanded

Fig. 4.5. The price elasticity of supply with respect to price.

characteristics which require that we modify the conventional form of the market model.

4.4 Modifying the Conventional Form of the Market Model

4.4.1 Time period

At the start of this chapter we defined demand as the quantity of a good that consumers are able and willing to buy at different prices over a specified period of time, other things remaining constant. We defined supply as the quantity of a good that producers are able and willing to offer to the market at different prices over a specified period of time, other things remaining constant.

The period of time that is relevant to this book is the very short term – in the case of grain and other agricultural commodities, this means one production cycle. For most field crops in temperate climates, this is one year – the crop being sown in the winter or spring and harvested in the summer or autumn. For most field crops in tropical zones, the production cycle is also one year – the crop being planted at the start of the rainy season and harvested during the dry season.

4.4.2 Demand for grain is inelastic

This book is concerned with supply and demand – but supply and demand for what? Farmers supply grain but consumers do not demand grain. They demand food, feeding stuffs and biofuels. In order to develop a model of the market for grain, we have to convert the products of grain back into grain itself. In economic terms, the difference between the products of grain and the grain itself is the cost of processing. Processing is the conversion of grain to useful products. The demand for the products of grain (e.g. bread, chipatis) is known as the 'primary demand'. The demand for the raw grain itself is known as the 'derived demand' because it is derived from the demand for the food product. The relationship between primary demand and derived demand is shown in Fig. 4.6.

The demand curve for the products of grain (e.g. food) is positioned above the demand curve for the raw grain itself. The gap between the two curves represents the cost of processing – that is, the cost of converting wheat to bread, and so forth. However, to maintain the readability of the figures in the rest of the book, we omit the demand curve for the products of grain. We include only the derived demand curve for grain itself. Our model therefore consists of a supply curve for grain and a demand curve for grain – supply and demand concern exactly the same good. We now need to know the slope of the curve that describes the derived demand for grain. Is this horizontal, vertical or somewhere between the two?

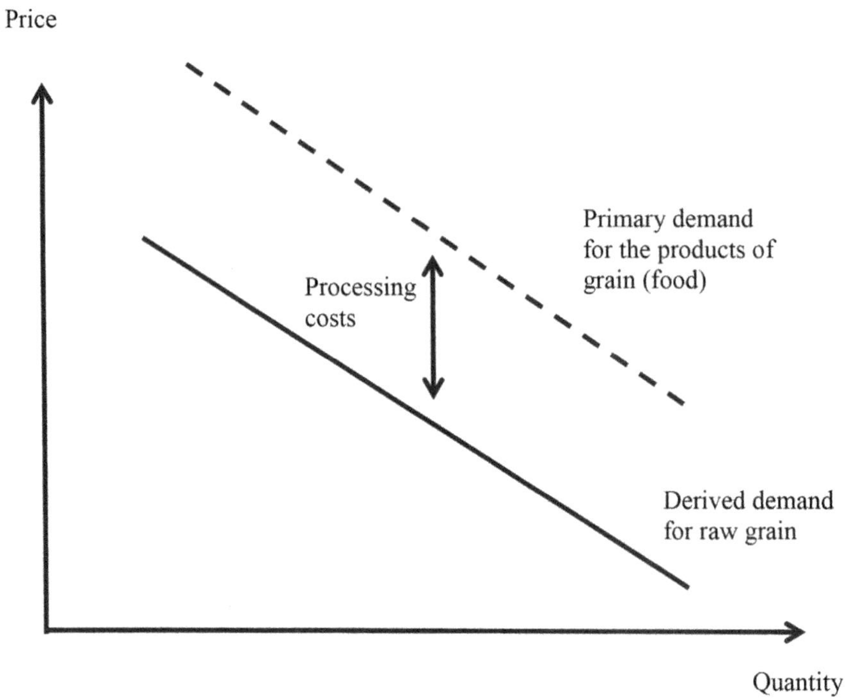

Fig. 4.6. Primary demand and derived demand for grain.

The slope of the derived demand curve relative to the slope of the primary demand curve can be explored using a technique known as marketing margin analysis. If the marketing margin is constant, that is, if the mark-up in price remains the same irrespective of the quantity being marketed, then the slopes will be the same. In general, it is reasonable to infer that a steep (inelastic) primary demand curve will imply also a steep (inelastic) derived demand curve.[2]

The next stage is therefore to ask: What is the elasticity of demand for food? Demand for food tends to be inelastic with respect to price. The quantity of food that consumers demand tends to change very little with price. This is the case whether prices are rising to a higher level or falling to a lower level.

A rising market: If the price of grain increases, consumers tend to buy about the same quantity as before. They tend not to cut down very much on their consumption of grain-based food products. This is because grain is essential to life. If root crops – such as potatoes and cassava – are available, some consumers may switch to them as they are usually cheaper than grain. But, generally speaking, consumers have little choice but to continue to buy about the same quantity of grain, irrespective of how expensive it is. To employ the economic terminology which we introduced in the previous section, when prices are rising, the demand for grain is inelastic with respect to price. This is an ineluctable feature of grain, true the world over and at all times in human history. Only when the price of grain is unaffordable will consumers give up buying it. At this point there is no economic demand for grain. When a person cannot afford to buy grain he cannot express any economic demand for it.

A falling market: How do consumers react when grain becomes cheaper? Do they buy more grain-based food products or do they buy less? In the case of goods such as clothes, electronic appliances, books and so forth, the cheaper they become, the more consumers tend to buy. But this is not so with grain.

In a falling grain market, there is little or no increase in the quantity of grain that consumers buy. However low the price, consumers buy, more or less, the same quantity. Quantity does not change appreciably, for the simple reason that each person has only one stomach and each stomach is of fixed dimension. The quantity of grain that each person can eat is anatomically constrained. Consumers cannot eat more than their stomachs permit.

We see that in both situations – whether prices are rising or whether they are falling – the quantity of food that is demanded changes very little. It means that across the whole range of prices – from high prices to low prices – the demand for food is inelastic with respect to price. This is the case over all time periods – a week, a month, a year, the short run and the long run.

Food is the main product from grain. The other products are beverages, feedstuffs and biofuels. Is the demand for these also inelastic with respect to price? The demand for grain for beverages and feedstuffs tends to be more elastic than for food, since the beverage manufacturers and livestock farmers are not willing to buy the same quantity if prices rise to very high levels. At high prices, they cut back on the quantity that they buy. As for the demand for grain for biofuel, this is inelastic if governments oblige fuel companies to add a certain quantity of biofuels to gasoline. Some governments do this in an attempt to switch drivers from fossil fuels to biofuels and thus reduce net emissions of carbon dioxide. To conclude this section, we consider that the aggregate demand for grain is inelastic with respect to price. In other words, if the price of grain changes a lot, the quantity demanded changes by a relatively small amount, and vice versa.

4.4.3 Supply of grain is perfectly inelastic

If the price of grain rises within the season, do farmers increase the quantity that they supply? If the price of grain falls within the season, do farmers decrease the quantity that they supply? If grain was an ordinary or general good – such as shirts, mobile phones or cars – we would expect this to be the case. A supply curve such as that in Fig. 4.1 would be a plausible description of the supply of grain.

It can be argued that, within a time period of a season (known as the very short run), the supply of grain is very different to the supply of other goods. It is not elastic with respect to price, as is the supply of many other goods. To the contrary, the supply of grain is perfectly *in*elastic with respect to price.

The reason is as follows. The conventional model shown in Fig. 4.1 describes how much grain farmers *plan* to produce on the basis of prices that they *expect* to receive when they sell their goods after harvest. If they expect to receive *higher* prices when they harvest their grain, they generally will plant a larger area and thus supply a bigger quantity. If they expect to receive *lower* prices when they harvest their grain, they will tend to plant a smaller area and thus supply a smaller quantity. In other words, the price that farmers expect to receive in the future – after harvest – influences the quantity of crops that they intend to supply to the market.

In this book we are not considering the quantity that farmers *plan* to produce at harvest time; we are considering the quantity of crops – in our case, grain – that farmers *actually* produce. This is described not by a supply curve which slopes up and to the right but by a supply curve which is vertical and positioned on a point on the *x*-axis which corresponds to the quantity of grain that has been harvested. Such a supply curve describes the supply that has *actually* occurred at the harvest. Whether the price at harvest time is high or low, farmers cannot influence the quantity of grain that is on the market because the harvest has already happened. The process of production has now finished for that particular season. This is the reason why the supply curve of grain, in the very short run, is a vertical straight line. It is not a line that slopes upwards and to the right, as is the case with ordinary and general goods, and illustrated in Fig. 4.1. [3]

Figure 4.7 shows the model of the market for grain, taking into account the particularities of grain supply and demand in the very short run. The demand curve is inelastic and represents the quantity of grain that consumers are willing and able to buy over a range of prices. The supply curve is perfectly inelastic and intersects the quantity axis at the level of the harvest.

4.5 Using the Model to Explain the Volatility of Grain Prices

4.5.1 Shifts in supply

Let us now explore what happens when there are shifts, from one year to the next, in supply and demand. The supply of grain onto the market is the quantity of grain that farmers harvest and sell. The quantity of grain that farmers harvest (the size of the harvest) is the product of two factors: the number of hectares that farmers have planted and the yield of grain from each hectare. The number of hectares planted to grain is decided by each farmer. It is limited by the size of the farm and usually depends on such factors as the labour

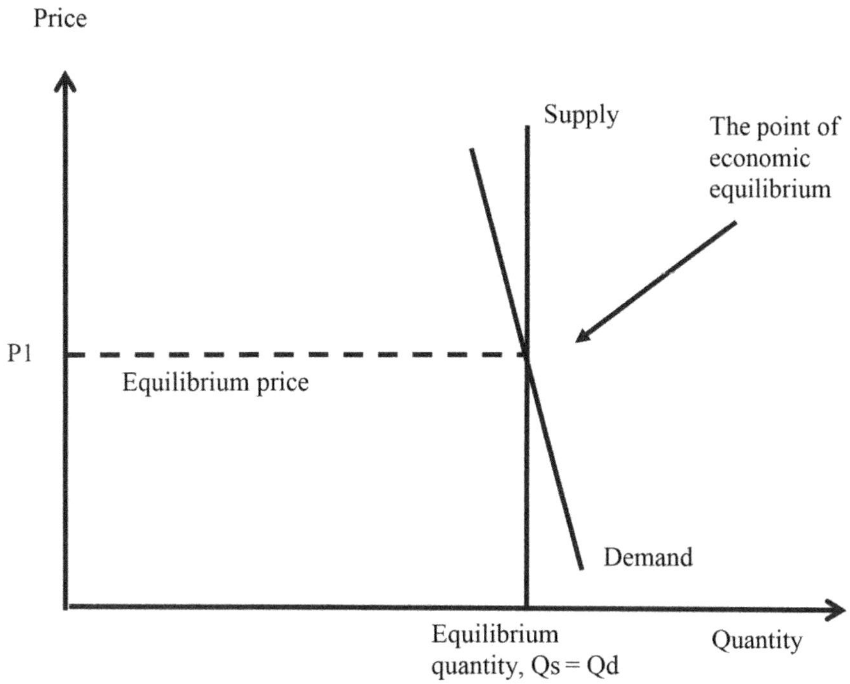

Fig. 4.7. An economic model of the grain market within the season.

and machinery that the farmer has available and the relative financial attractiveness of growing grain compared to alternative crops. The yield of grain from each hectare depends on the amount of labour that is available during the growing season, the farmer's agronomic methods, the quantity of inputs (such as irrigation water, fertilizer) and on the beneficence, or otherwise, of nature during the growing season. Nature can bring crop diseases, crop pests and bad weather, in particular too much or too little rain.

In some circumstances, farmers may have a degree of control over crop pests and diseases. In other cases – such as when they cannot afford to buy pesticides or hire extra labour – pests and diseases can result in big reductions in crop yields. As for the weather, farmers can do very little to counter the effects of bad weather (although irrigation can mitigate against a drought). While very good weather can result in high yields, bad weather – particularly lack of rainfall – can result in very low yields.

Figure 4.8 shows the effect of an expansion of supply, caused by a good harvest. Supply expands from S1 to S2. The expansion of supply, coupled with the inelastic demand, combine to cause prices to plunge from P1 to P2. The price of grain falls a lot, even though the supply increases by relatively little.

Figure 4.9 shows the opposite situation – a contraction of supply. This is due to a bad harvest. Bad harvests are usually caused by poor weather, such as late frosts, floods or droughts, during the growing season or at the

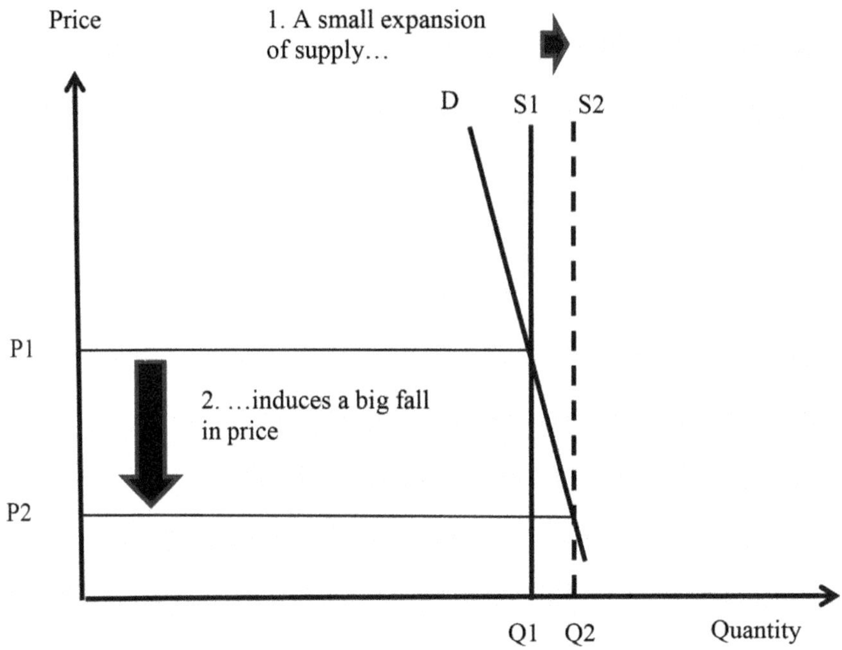

Fig. 4.8. A small expansion of supply induces a big fall in price.

time of the harvest. A bad harvest need not only be caused by poor weather; it could, at least in principle, also arise if there are not enough people to plant, weed and harvest the grain. In other words, a bad harvest can be due to a lack of farm labour. This can be the consequence of a national epidemic or global pandemic when governments impose restrictions on people's travel. Travel restrictions were one of the measures taken by many governments during the Covid-19 pandemic in early 2020.

Figure 4.9 depicts a bad harvest as a leftward shift of supply from S1 to S2. What is the effect on price? The price rises from P1 to P2. The price rises proportionately more than the quantity falls. The reason for the disproportionate rise in price is the inelastic demand – consumers have to eat and are willing to pay a high price for grain in order to secure their minimal dietary requirements. A small shift in supply induces a big change in price.

4.5.2 Shifts in demand

The demand for grain can also fluctuate. Fluctuations in demand are, however, likely to be less dramatic than fluctuations in supply. If there is a shift in demand, this is likely to be gradual and spread over several seasons, rather than occurring from one season to the next. The gradual nature of shifts in demand allows farmers to respond to them by planning to produce more or

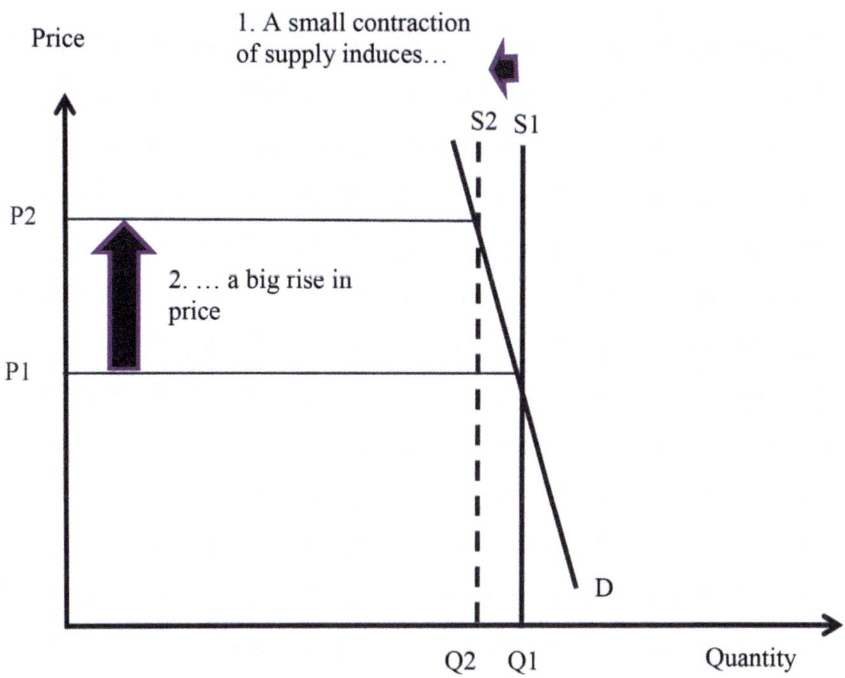

Fig. 4.9. A small contraction of supply induces a big rise in price.

less grain as the case may be. Consequently, the main causes of sudden rises and falls in grain prices are shifts in supply, rather than shifts in demand.

However, a small short-term expansion of demand would cause a big increase in price. This is shown in Fig. 4.10. Demand expands from D1 to D2. Since supply is fixed at the size of the harvest, the new equilibrium is established at the same quantity, Q1, but at a higher price. The price has risen from P1 to P2.

A contraction of demand – that is, consumers demanding less grain than in the previous time period – causes a relatively big drop in price. This is shown in Fig. 4.11. Demand contracts from D1 to D2. Since supply is fixed at the size of the harvest, the new equilibrium is established at the same quantity, Q1, but at a lower price. The price falls from P1 to P2. The fall in price reflects the fact that although there is less demand for grain, there is still the same quantity on the market for sale.

4.5.3 Simultaneous shifts in supply and demand

Of course, supply and demand may shift at the same time. It could be that there is a poor harvest (a contraction in supply) at the same time as demand is expanding. In this case the price escalates dramatically, from P1 to P2, as shown in Fig. 4.12.

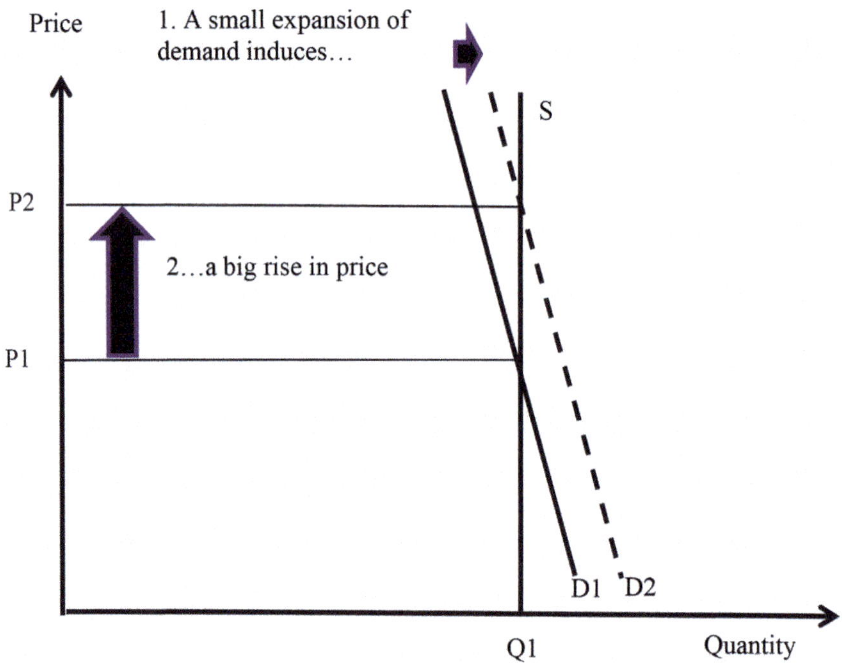

Fig. 4.10. A small expansion of demand induces a big rise in price.

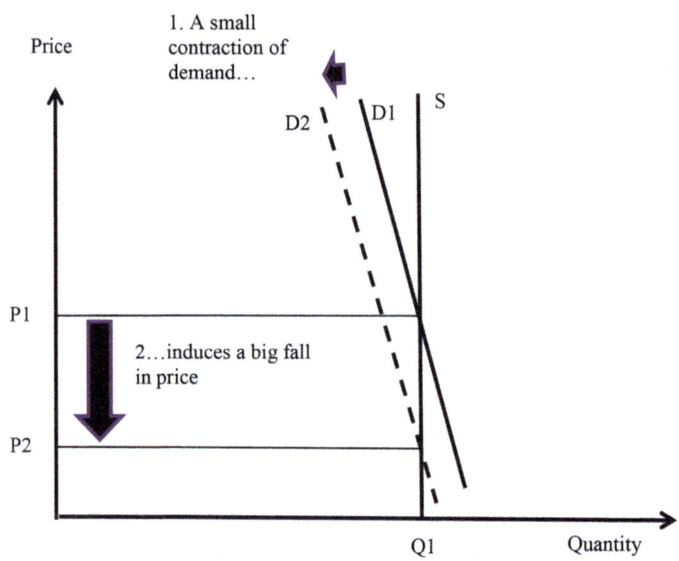

Fig. 4.11. A small contraction of demand induces a big fall in price.

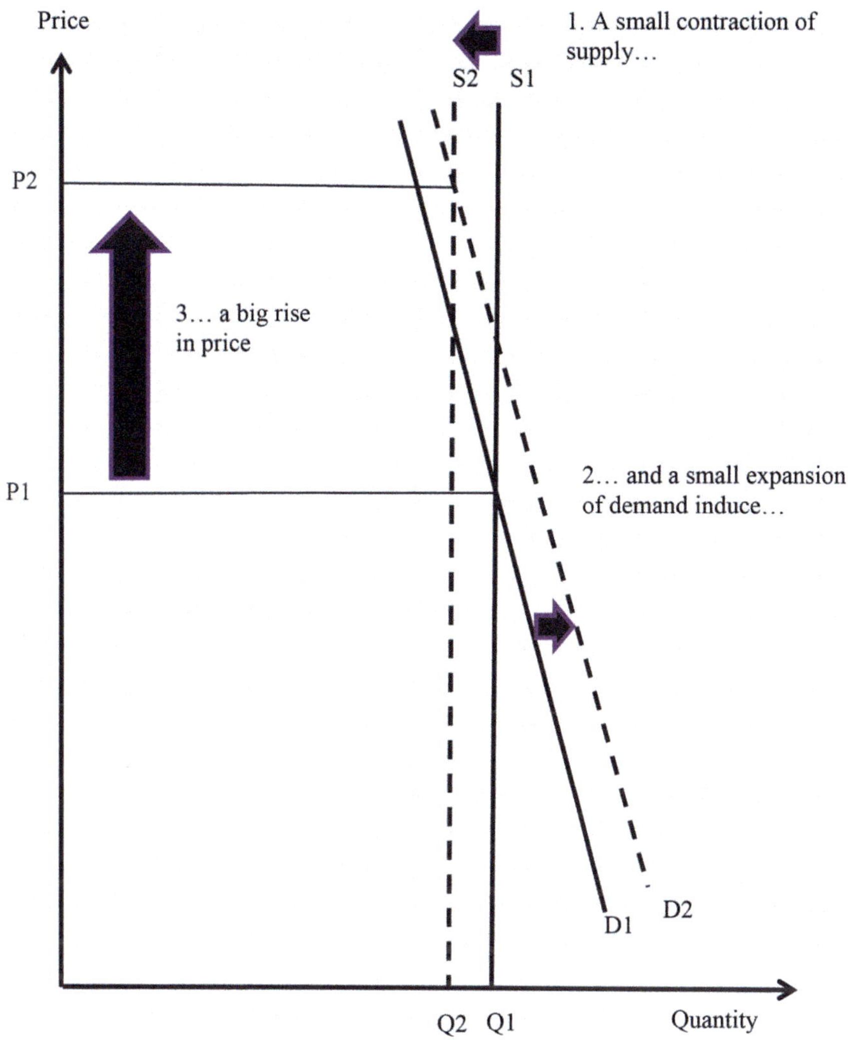

Fig. 4.12. A simultaneous expansion of demand and contraction of supply.

By the same token the price drops dramatically, from P1 to P2, if there is a simultaneous expansion of supply and contraction of demand, as shown in Fig. 4.13.

The final possibility, which is not shown, is that a shift of supply cancels out a shift in demand. Suppose, for instance, there is a bountiful harvest (an expansion of supply) in the same year that demand for grain expands. In that case, both supply and demand shift to the right. The quantity of grain supplied and the quantity of grain demanded increase in tandem. Because both supply and demand expand by the same extent, the price remains the same.

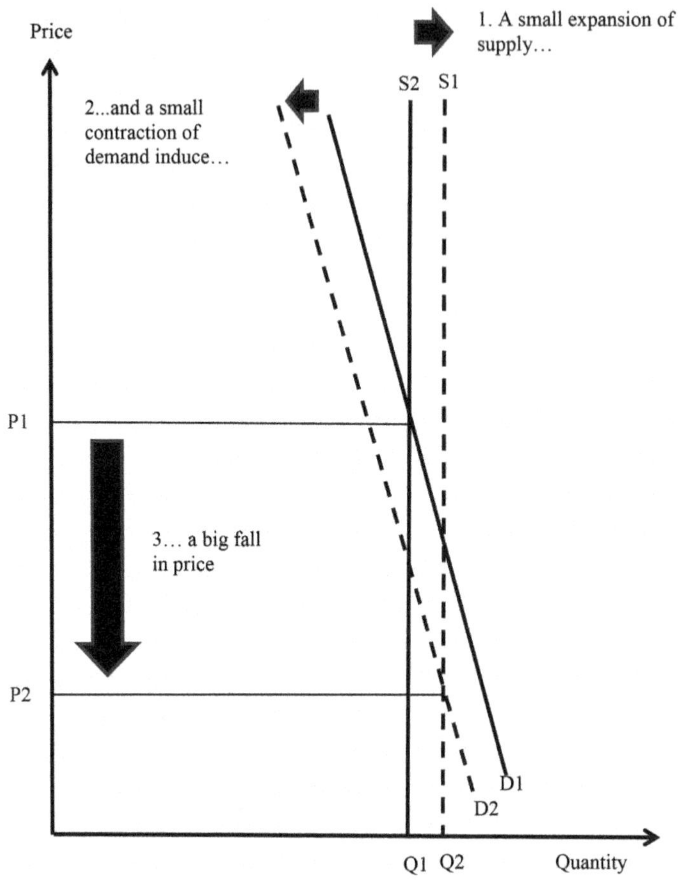

Fig. 4.13. A simultaneous expansion of supply and contraction of demand.

Alternatively, suppose that the harvest is meagre in the same year that incomes fall. A meagre harvest means that supply contracts. A fall in incomes means that demand also contracts. Both supply and demand will shift to the left. The quantity of grain that is supplied and demanded falls, but the price remains the same. The contraction of supply has been cancelled out by the simultaneous contraction in demand.

4.6 Is It Plausible to Prevent Shifts in Supply and Demand?

Would it not greatly simplify things if there were no shifts of supply and demand? If supply and demand would remain the same, year after year, prices would be stable. There would be no volatility problem. We would have no problem of food becoming too expensive or farmers facing a profit wipe-out.

To all intents and purposes, it is implausible to prevent shifts in the supply and demand for grain. On the supply side, this would mean keeping the same weather and growing conditions year after year. It would also require farmers to plant the same number of hectares of grain every year. This is well nigh impossible. First, it would require the government to very closely supervise the actions of farmers, which would be costly from an administrative point of view. Secondly, farmers would likely resist this high level of control on the grounds that the government was overly interfering in their farming decisions.

As for preventing shifts in demand, people's incomes and tastes would have to remain the same year after year, a prospect which is clearly implausible. It would also mean stopping any increase or decrease in human population.

Not only is it implausible to prevent shifts in the supply of and demand for grain, it is also undesirable from a broader perspective. Prices are economic signals – they embody vital information. The market sends this information to all who participate in it. Through the medium of market prices, farmers know whether consumers want more or less grain, whether they want grain of a different type and grade, etc. Farmers can adjust their farming decisions in the light of such information. Likewise, prices convey essential information from farmers to consumers, such as which grains are in plentiful supply and which grains are scarce. Consumers adjust their consumption patterns accordingly. Prices are thus the conduit by which farmers and consumers converse with each other on all matters of mutual concern. Freezing supply and demand might solve the problem of short-term volatility but it would eliminate the dialogue between farmers and consumers. Adam Smith's hidden hand would no longer be guiding and co-ordinating farmers and consumers. For instance, in some years farmers could produce far too much grain and in others far too little. Farmers would have no means of learning about changes in the preferences and tastes of consumers.

4.7 Conclusion

Grain prices are volatile because:

- the supply of grain is not fixed – it can expand and contract from one season to another;
- the demand for grain is not fixed – it can expand and contract over time;
- the supply of grain is perfectly inelastic with respect to price within the season; and
- the demand for grain is inelastic with respect to price.

The corollary is that a small shift in supply or demand can induce a large change in price. This is the root cause of the volatility of grain prices.

But is volatility a bad thing? Does it not merely reflect the natural workings of the market? The following chapter considers the winners, losers and arguments for and against more stable grain prices.

Notes

[1] Demand is said to be elastic when the proportionate change in quantity is greater than the proportionate change in price, and inelastic when the proportionate change in price is greater than the proportionate change in quantity. This means that when the demand 'curve' is a straight line, the response of quantity demanded to a change in price will become progressively more elastic as the price increases and the associated quantity decreases. However, if the demand curve has a steep slope, then there will be an inelastic response over the majority of the price range covered. It is conventional to describe this as an inelastic demand curve.

[2] Ritson states that, applying marketing margin analysis, it can be shown that the derived demand facing the farm sector may be more or less price elastic than consumer demand for the food product, depending on the behaviour and structure of the marketing sector (see Ritson, 1977, chapter 3). In a personal communication of 14 May 2019, Ritson has elaborated on this: 'Where the vast majority of the farm product (say wheat) is used for the same consumer food product (say bread), then the analysis is fairly straightforward. Things become much more complicated when the grain is used in the manufacture of a variety of food products (and some cereal finds its way into a surprising range of food products). The issue becomes even more complicated when the use of grain for animal feed is considered. First, the demand for the farm livestock product must be derived from the consumer demand for the meat/milk/dairy product/eggs; and then the demand for the animal feed has to be derived as a farm input from the demand for the farm livestock product. The most straightforward example is probably eggs, where the farm product receives limited 'marketing' (grading, packaging, transport and retailing); and the cost of animal feed represents a high proportion of the overall cost of egg production.'

[3] Ritson (1977, p. 109) points out that 'the supply of an agricultural product is sometimes described as perfectly inelastic in the very short run. The 'very short run' can be thought of as a period of time which does not encompass a full crop year....Some elasticity of supply in the very short run can be introduced by the accumulation or running down of stocks but only if it is the practice to hold 'carry-over' or 'year-to-year' stocks. Otherwise, farmers (and others) holding stocks from the most recent harvest can influence the timing of supply throughout the very short run, but not the total quantity supplied during that period.'

Reference

Ritson, C. (1977) *Agricultural Economics: Principles and Policy*. Crosby Lockwood Staples, UK.

5 Unstable Grain Prices – Who Wins? Who Loses?

5.1 Introduction

The previous chapter explained how shifts in supply and demand can cause big changes in price. Shifts in supply and demand for grain lie at the heart of price volatility. The present chapter asks: Does volatility matter? Do people suffer? Is society harmed in some way? If so, how?

We start by defining 'farmers' (5.2) and 'consumers' (5.3). In Section 5.4, we discuss the four circumstances that can cause volatile prices: an expansion of supply of grain, an expansion of demand, a contraction of supply, and a contraction of demand. Each circumstance has a real-life effect – either beneficial or adverse – on farmers and consumers. They are summarized in Section 5.5. We ask what are the effects of price volatility on governments (5.6) and on society as a whole (5.7).

5.2 Who Do We Mean by Farmers?

There are two types of farmer in the world: subsistence farmers and commercial farmers. In this book, we are not concerned with subsistence farmers as such, because they are not part of a market. They grow grain for their families, not for sale. They are not affected by rising or falling grain prices. It is important to note a nuance, however. Consider that, for some reason, such as a drought, subsistence farmers do not manage to grow enough to feed their families. They are then forced by circumstance to buy their food from the market. They are no longer subsistence farmers; they are now consumers.[1]

In this book, we are concerned with commercial farmers. These are farmers who grow grain for the market. They earn money by growing and selling grain. It may be their sole crop or one of half a dozen crops. Commercial farmers are directly and immediately affected by the instability of grain prices.

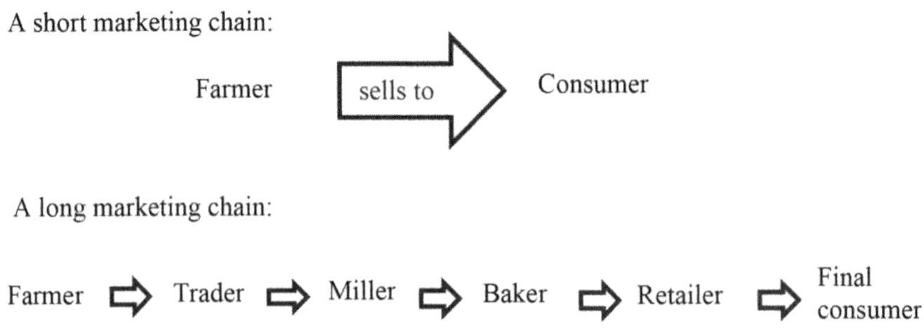

Fig. 5.1. Short and long marketing chains.

5.3 Who Do We Mean by Consumers?

As we noted in Chapter 2, grain is not only used for human food; it is also used for industrial purposes, biofuels and animal feed. The term 'consumer' refers to all those people who buy the products that are derived from grain. These are foodstuffs, fuels, industrial materials and feedstuffs. Consumers are linked to farmers by a marketing chain, the length of which determines the extent to which consumers 'feel' the effects of price volatility. Two examples of a marketing chain – one short, the other long – are shown in Fig. 5.1.

The short marketing chain is illustrated by a farmer in, say, Africa who grows maize, takes it to the market and sells it directly to the consumer. As a second example, we could cite a farmer in Europe who produces milk, turns the milk into 'farmhouse cheese' and sells it directly to the consumer in a farmers' market. In both cases, only two parties are involved – the farmer and the consumer. There are no intermediary middlemen or processors. In a short marketing chain, the full impact of any change in price falls on the consumer. There are no intermediate processing stages that can dilute the impact of a rise in the price of the raw material.

In contrast, in a long marketing chain there are several, perhaps half a dozen, stages through which the grain passes between the farmer and the consumer. Various participants (traders, transporters, storers, millers, bakers, wholesalers, retailers and so forth) along the chain add value. They are remunerated for their services. The total remuneration of all the participants between the farmer and the consumer constitutes the processing costs that we discussed earlier in Section 4.4. At each processing stage the value of the good is increased, as Table 5.1 illustrates.

The table shows a marketing chain that transforms a bushel of grain – equivalent to around 30 kg – into loaves of bread. The original grain was worth $3. The loaves of bread are worth $30. The marketing chain has increased the value of the grain by a factor of ten.[2]

Table 5.1. A marketing chain, transforming grain into bread.

		At grain price of $3 per bushel	
Participant	Function	Price of function that the participant performs ($ per bushel)	Accumulative price ($ per bushel)
Farmer	Provides the grain	3	3
Trader	Transports and stores	5	8
Miller	Makes the flour	6	14
Baker	Bakes the bread loaf	7	21
Retailer	Markets the loaf	9	30
Final consumer	Buys and eats the loaf		

5.4 The Consequences of Price Volatility

5.4.1 A rise in price due to a contraction of supply

We now turn to the consequences of a rise in the price of grain. Figure 5.2 uses our model of the market to show a rise in price due to a contraction of supply.

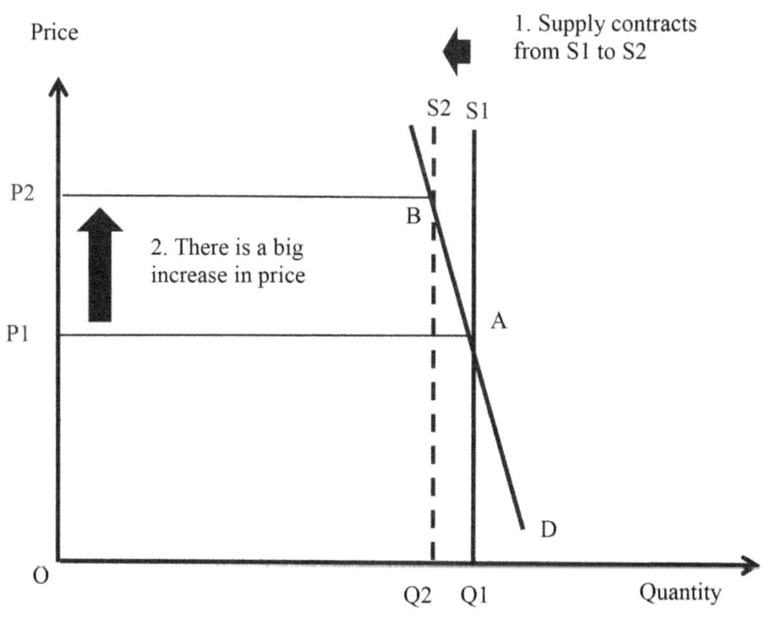

Fig. 5.2. A rise in price due to a contraction of supply.

5.4.1.1 Effect of a rise in price on consumers

Before the contraction of supply, consumers buy a quantity of grain equal to Q1. They pay a unit price of P1. The amount of money that consumers spend on grain before the contraction of supply is represented by the area OP1AQ1. After the supply has contracted, consumers buy a smaller quantity (Q2). Because demand is inelastic, the price per unit of grain has risen from P1 to P2. This means that the amount of money that consumers spend on grain is OP2BQ2. This area is larger than OP1AP2, indicating that consumers now spend more money for less grain.

Is this a problem? Do consumers suffer harm if they pay a higher price for their bread and rice and purchase less of it? If consumers are poor, an increase in price can affect them very badly. In the worst cases, they can suffer hunger and even die from starvation.

As an example of the problems wrought by an escalation of grain prices, we may remind ourselves of France in the 18th century. The average French worker spent half his daily wage on bread. But when the grain crops failed two years in a row, in 1788 and 1789, the cost of bread shot up to 88% of his wages (Neely, 2008). This caused food riots in Paris and helped to trigger the French Revolution. The entire political establishment was swept away with great cruelty and much spilling of blood. It is an example of the consequences that can befall a government if food becomes unaffordable to the people.

It is sometimes thought that poor consumers are found only in poor countries, but this is far from the case. There are many families in rich countries who are poor. They can be badly affected by an increase in the price of grain. A rise in the price of grain obliges poor families to spend more on food and therefore less on other goods and services. In the USA, the plight of poor families is alleviated by a food stamp programme (which is relatively costly for the US federal government). In other rich countries, poor families may receive charitable donations of food provided by food banks.

In contrast to poor consumers, rich consumers may not be unduly affected by an increase in grain prices. This is for two reasons: first, food, as a whole, constitutes a smaller portion of total household expenditure among rich families (15% or less) than it does among poor families (30% or more). Because of their bigger incomes, rich consumers can absorb a price increase more easily than poor consumers.

Secondly, rich consumers are more likely to be at the end of a long, rather than a short, marketing chain. The impact of a higher price for grain will be diluted by the various stages of processing and transport that occur in the chain as the grain is transformed into food. Table 5.2 shows how the price of wheat can double but cause only a 10% increase in the price of a loaf of bread.

Table 5.2. The dilution of a price rise in a long marketing chain.

		At grain price of $3 per bushel		At grain price of $6 per bushel – it has doubled in price	
Participant	Function	Price of function ($ per bushel)	Accumulative price ($ per bushel)	Price of function ($ per bushel)	Accumulative price ($ per bushel)
Farmer	Provides the grain	3	3	6	6
Trader	Transports and stores	5	8	5	11
Miller	Makes the flour	6	14	6	17
Baker	Bakes the bread loaf	7	21	7	24
Retailer	Markets the loaf	9	30	9	33
	Proportionate increase in the price of a loaf of bread due to a doubling of the price of grain, where the absolute mark-up remains the same				(33-30)/30 =10%

In poor countries, the marketing chain is usually short. In rich countries, the marketing chain is usually long. This is a reason why consumers in poor countries are hit proportionately more by a rise in the price of grain than consumers in rich countries. When the price of grain rises on the world market, the reactions of consumers in poor countries are generally much more vehement than those of consumers in rich countries.

5.4.1.2 *Effect of a rise in price on farmers*

As for farmers, they are ostensibly the winners in an increase in price due to a contraction of supply. Before the price increase, their revenue from sales was the same as the consumers' expenditure – OP1AQ1, in Fig. 5.2. After the shift in supply, their revenue from sales rises to OP2BQ2. This is the rationale behind the farmers' toast: 'Here's to the next drought or war'. Few good citizens would raise a toast to such a calamity. Why would farmers welcome malevolence? However, on occasion, farmers may do just this.

This may, at first sight, seem paradoxical. If there is drought, it means that the farmer produces less. If he produces less, then it follows that he has less to sell. If he has less to sell, how can he gain more money? Surely droughts are to be avoided at all costs. How is it that farmers may raise their tankards to toast the advent of a drought? The answer is that when farmers sell less grain – as happens when there is a drought – the price can shoot up. For instance, the farmer may find that, in a drought year, instead of selling ten bags of rice, he has only eight bags to sell (a fall of 20%). But he can expect the price to rise from $10 a bushel to $15. This is a price increase of 50%. The fall in the quantity harvested is more than offset by the rise in price. Overall, he will be better off. Paradoxical though it may seem, farmers can be better off when there is a drought.

Of course, the drought could be very severe and all crops could fail. Farmers then have nothing to sell and cannot benefit from the higher prices. But in years of mild drought – or of drought in another area of the country – the economics of agriculture work in the farmer's favour.

Nevertheless, as is so often the case, a closer analysis finds the situation to be somewhat nuanced. It is wrong to assume that when the grain price rises, all farmers fill their pockets with windfall profits. This is not necessarily the case. A farmer will have to purchase his seed, fertilizer and chemical pesticides from an agricultural supplier. He may sell his grain to a merchant. He may rent his land from a landowner. He may have to renew his farm equipment with a machinery dealer. These suppliers and merchants often keep a close watch on the financial situation of the farmer – the farmer is, after all, their client and customer.

When there is an increase in the price of grain, these suppliers and dealers may capture the higher prices for themselves. For instance, in the following season, farmers may find that the prices for inputs rise because all the merchants take advantage of the fact that farmers have made good profits. At the next review of the rent, the landowner will press the case that since farming has become more profitable, isn't it only fair that there is an increase in rent?

This is to say that, when grain prices rise, farmers may not keep all the increase for themselves. It is likely that some of the benefit will be extracted and captured by the people who supply farm inputs and market farm products. To sum up, when a contraction of supply forces the price of grain to increase, consumers always lose out. They lose out because food costs more. If the consumers are poor, they may suffer greatly from hunger and may even die from starvation. If the consumers are rich they may be able to absorb the price increase without a huge problem. As for farmers, ostensibly the opposite applies. In principle, they should end up as the winners. After all, they get higher prices for their grain. But farmers only win if two conditions apply: they have grain to sell; and their suppliers and landowners do not siphon off the windfall gains for themselves.

5.4.2 A rise in price due to an expansion of demand

Such a situation is shown in Fig. 5.3.

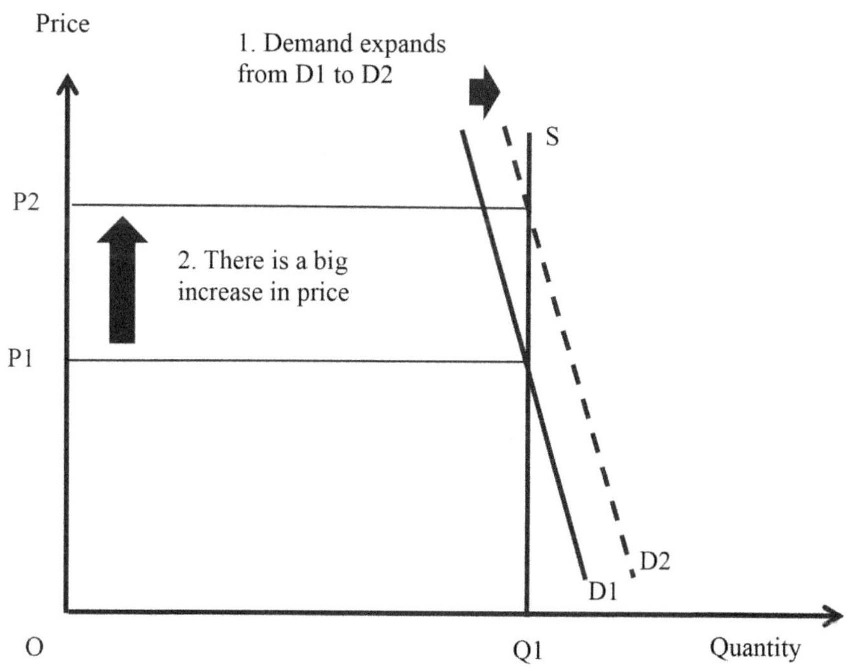

Fig. 5.3. A rise in price due to an expansion of demand.

5.4.2.1 *Effect of a rise in price on consumers*

In this situation, the quantity of grain that farmers offer on the market does not change – it remains at Q1. The expansion of demand has caused the price to rise from P1 to P2. Consumers are faced with higher food prices. They are the losers. As above, poor consumers may suffer hunger, even starvation. Rich consumers may be able to absorb the higher prices.

5.4.2.2 *Effect of a rise in price on farmers*

Farmers are the winners in this situation. They sell the same quantity of grain as before (unlike in the situation in which supply contracted) and the price that they receive has risen from P1 to P2.

However, the same caveat applies: farmers only gain if their landowners and suppliers do not siphon off the windfall gains for themselves.

5.4.3 A fall in price due to an expansion of supply

This situation is shown in Fig. 5.4.

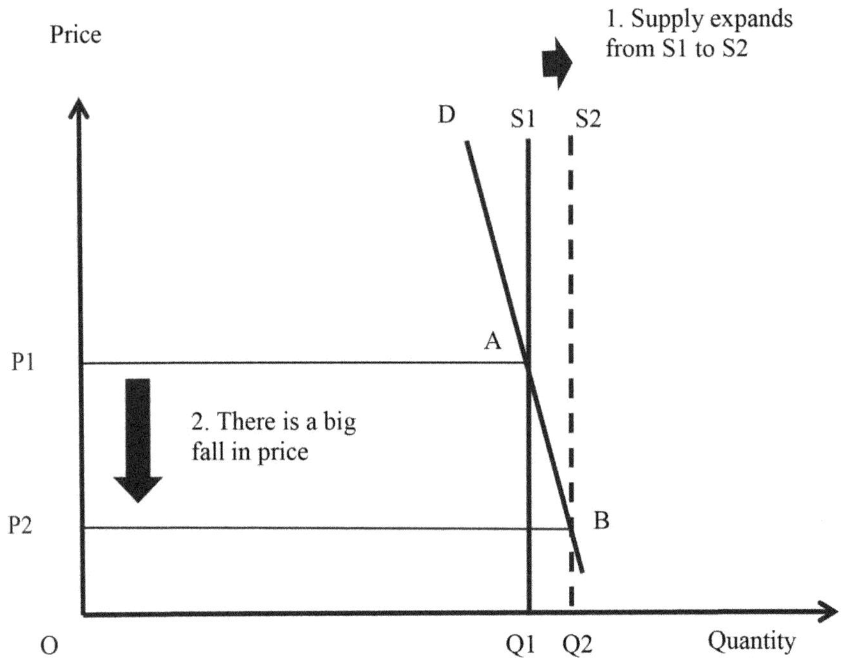

Fig. 5.4. A price fall due to an expansion of supply.

5.4.3.1 *Effect of a fall in price on consumers*

In this situation, the supply of grain has expanded from S1 to S2. This has caused the price to fall from P1 to P2. Before the fall in price, consumers spend an amount of money that is equal to the quantity of grain that they bought multiplied by the price per unit. The quantity was Q1. The price per unit was P1. Therefore, the amount of money that consumers spent before the fall in price is represented by the area OP1AP2.

After the expansion of supply, the quantity of grain bought by consumers increases from Q1 to Q2. This means that consumers are better off – they consume more grain (to be precise more foods derived from grain). Consumers are also better off due to a second effect. Clearly, the price per unit of grain is now less – it has fallen from P1 to P2. There is, however, an effect on total expenditure on grain. The total amount of money that consumers spend on grain is now less. The amount of money that consumers spend on grain is represented by the area OP2BQ2. This area is less than the area OP1AQ1. The reason that consumers spend less money and yet consume more grain is because the amount by which the price of grain has fallen is greater than the amount by which the quantity consumed has increased. This reflects the fact that demand for grain is inelastic with respect to price. The upshot is that consumers are better off in two ways: they consume more grain (and grain-based foods) and they have more money to spend on other goods.

However, the situation is not always quite as simple as it may first appear. This is because of the food chain. Let us say that the food chain consists of only the farmer and the consumer – with no intermediaries such as millers, bakers and supermarkets. The farmer takes his grain to the local village market. He sells it to the consumer who will grind it himself or herself and cook the family's meal. This is a direct transaction between the farmer and the consumer. In this case, the consumer captures the full benefit of the low price.

Many consumers, however, in rich countries are at the end of a food chain. Very few buy directly from the farmer. The food chain may consist of three, four or more stages. The first firm in the chain – most probably a trader – will buy the grain at the lower price. The question then arises: Does this first participant pass the lower price down the chain to the consumer or does he keep the lower price for himself and sell the product on to the next participant at the previous higher price?

It depends on market power. Those firms with power in the marketing chain tend to keep the benefit of low prices for themselves. They do not pass it down to the next participant. The individual consumer is weaker that the other participants. Indeed, consumers as a whole are weaker than the other participants because consumers tend to act atomistically. They are not organized, nor do they act in concert with each other. Because of this they are weak *vis-à-vis* the other participants in the chain.

Having said this, it is usually the case that merchants, trader, bakers, millers and all the other participants in the food chain are in competition with other merchants, traders, bakers and so on. The degree to which a participant keeps the benefit of a fall in grain prices to himself has to be tempered by the degree to which the participant is in competition with others. After all, the participants want to keep their clients contented and satisfied. Otherwise they may forfeit market share.

To sum up, when the price of grain falls, consumers are – at least in principle – the winners. They buy more food and they spend less money on food, allowing them to spend more money on other goods. But the extent to which they capture these benefits depends on the degree to which the other participants in the chain pass on the lower prices.

5.4.3.2 *Effect of a fall in price on farmers*

In stark contrast to consumers, who win when there is a fall in the price of grain, farmers lose out. When the price of grain falls, say due to a good harvest, farmers are worse off. Although they sell more (the quantity that they sell increases from Q1 to Q2), the fall in price (from P1 to P2) is proportionately greater than the increase in the quantity sold. It is the consequence of demand for grain being inelastic with respect to price.

The slump in prices that can accompany a good harvest lies behind a poignant inscription (see Fig. 5.5) chiselled on a tombstone in the 19th century in a cemetery in Saskatchewan, Canada.

These lines encapsulate the tragedy of the oversupply of grain – and all other farm products. It epitomizes the plight of farmers. Just as it seems strange for farmers to raise their glasses to a drought, it seems even more bizarre that farmers can starve when the harvest is plentiful.[3] We saw that a drought can push up prices more than they cause yields to fall. But it was the opposite situation that befell Farmer Pete. The harvest was exceptionally good; large quantities of grain flooded the market, and the price fell like a stone.

Does a slump in farm prices automatically mean that farmers face the prospect of penury and starvation? After all, a slump in farm prices usually arises from a bountiful harvest – the farmers have plenty of sacks of wheat and bags of maize. They are not short of food to eat. They may complain of the low price and may lament the disappearance of their profits. But they surely cannot complain that their own kitchen cupboards are bare?

If farmers have debts from which they are unable to escape, they can fall into the situation of Farmer Pete. One can envisage Farmer Pete having to borrow money at the beginning of the season to pay for a horse to plough his fields, for seeds to sow, perhaps even for timber to fence his land and to build his hut. He was planning to repay his debts at harvest when he collected the money for his crop. In a year of normal harvest and normal prices, he would perhaps have had to sell half his harvest to pay his debts.

Fig. 5.5. The paradox of plenty. Farmer Pete was the victim of low grain prices. When there is a good harvest, the market can be flooded with grain. Prices fall like a stone. Farmers have no choice but to accept very low prices for their grain. Like many farmers, then and now, Pete may have had to borrow money to pay for his crop inputs (seed, labour, water). In this particular year, there would have been a good harvest, causing prices to collapse. Farmer Pete would have been obliged to sell all his grain to repay his debts, leaving nothing for himself and his family. In this way he starved to death. This is the paradox of plenty – what at first sight appears to be the beneficence of nature can be a curse for farmers.

But with prices at rock-bottom and without creditors willing to reschedule his debts, Farmer Pete would have been obliged to sell all his wheat at the prevailing rock-bottom prices to scrape together as much money as possible to repay what he owed. This would have left him with nothing to eat.

A second graphic illustration of the hardship that befalls farmers when prices slump has been provided by the author John Steinbeck. His novel *The Grapes of Wrath* recounts the story of farmers in the southern states of the USA during the depression of the 1920s (Steinbeck, 1939). The prices of grain and other agricultural commodities were at rock-bottom levels for several consecutive years. Farmers could not make any money. Thousands of farm families had no choice but to leave their farms, trekking to California in search of economic salvation. It was following this massive social tragedy that the USA first began to offer its farmers some relief from

Fig. 5.6. Food price volatility is a problem for all countries – prosperous ones included. This photograph was taken in 2019 in Ireland. This farmer – together with hundreds of others – was protesting against low prices for their products. The prices were low because there was a surplus of farm commodities on the market. Irish farmers were unable to make a profit from their work. Low prices could have been prevented if the government had kept the market price within a price band by means of a buffer stock. (Photo: REUTERS/Lorraine O'Sullivan)

the downside of price instability in the form of programmes to regulate the prices of farm commodities.

If the grain price is too low to cover the cost of production, the farmer is in a potentially precarious situation. After all, farms are businesses and when a business no longer makes a profit it is no longer viable. This is not just a problem for farmers in poor countries: farmers in rich countries are also vulnerable to falls in price. They are often on tight margins. If they cannot sell their products at a profit they will eventually give up farming, often having accumulated debts in a vain attempt to 'ride out' the bad years. Apart from the distress this causes to families and communities, countries can ill afford to lose their farmers because this can jeopardize the country's food supply (Fig. 5.6).

Farmers have a variety of mechanisms to cope with years of low prices. They spend less on non-essential items, they do not renew machinery, they draw on savings, they look for jobs and income away from the farm, they sell some land or livestock and they delay making payments to their creditors. Coping mechanisms can only last for so long. If a year of low prices is followed by one or several more years of low prices, farmers will be faced with the spectre, or indeed the reality, of bankruptcy and of having to leave the land and their homes to search for a livelihood elsewhere.

When prices fall to a low level, farmers often take to the streets to protest and demonstrate. They are expressing their anger and frustration. They are defending their families, their businesses and their livelihoods. It is no surprise that these demonstrations can become violent .

In other situations, farmers do not join together and protest publicly. Instead, they remain at home, becoming withdrawn and depressed as they sink further into debt. For some, the only relief is to take their own lives.

5.4.4 A fall in price due to a contraction of demand

This situation is shown in Fig. 5.7.

5.4.4.1 Effect of a fall in price on consumers

In this situation, the price of grain has fallen because demand has contracted. Consumers buy the same quantity of food as before (Q1) but they pay a lower price. Ostensibly, they are better off. However, firms at the front of the marketing chain may be in a position to keep the windfall

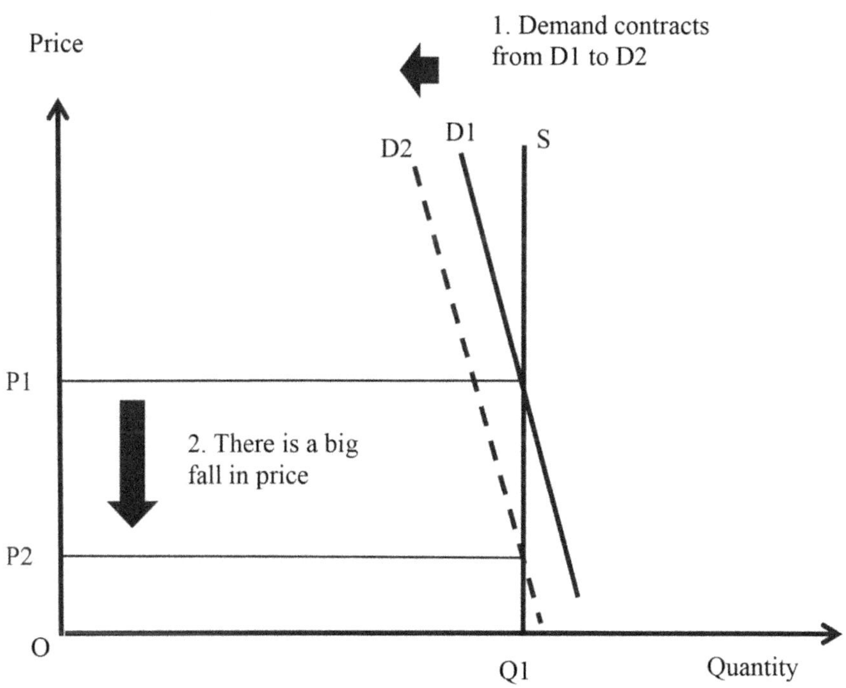

Fig. 5.7. A price fall due to a contraction of demand.

profits for themselves, rather than pass them on down the chain to the final consumers.

5.4.4.2 *Effect of a fall in price on farmers*

When the price of grain falls, farmers are worse off. They sell the same quantity (Q1) but the price falls from P1 to P2. The price may fall so low that farmers are unable to cover their costs of production and may find themselves facing financial difficulties, as in the case described above.

5.4.5 The human consequences of price instability

We have considered who gains and who loses from high and low prices. The simple picture is that consumers gain in those years when prices are low and lose in those years when prices are high. However, there is no symmetry – to have the benefit of cheap food in one year hardly compensates for hunger and malnutrition in the following year. A similar asymmetry confronts farmers – they may have good profits one year but these are no compensation for having to leave the land when prices plummet. Price volatility is not a zero-sum game for farmers and consumers. Table 5.3 provides a summary of the effects.

Table 5.3. The human consequences of unstable grain prices.

	Years of low prices	Years of high prices	Caveat
Consumers	Consumers benefit; food is cheap and consumers can buy food at low prices, providing the low prices are passed down the marketing chain.	Consumers lose out; food is expensive. In *extremis*, consumers may not be able to afford enough food. This can lead to hunger, even death by starvation.	Cheap food is no compensation for subsequent unaffordable food, with the prospect of hunger and starvation.
Farmers	Farmers lose out; grain is cheap. Farmers may not make sufficient profit to cover their costs. In extremis, they may become bankrupt and have to give up farming.	Farmers benefit; grain is expensive and farmers can sell it at a high price. Their sales revenue is high. Other players in the food chain may cream off some of the profits. Some farmers may have no grain to sell due to a failure of their harvest.	A year of high profits does not compensate a farmer for a future year of possible bankruptcy.

5.5 Adverse Effects on Governments

The above discussion has focused on individual citizens: farmers and consumers. What about the effects of the instability of grain prices on governments and on broader society?

5.5.1 Government disruption

Unstable prices are a disruption for governments. When grain prices are stable, governments have time and space to plan and implement social programmes – building roads, hospitals and schools, for example. Stable prices allow governments to focus their efforts and creative energies on long-term tasks rather than being distracted by emergencies.

In contrast, when prices are unstable, governments may have to abandon, or at least suspend, long-term programmes and plans. Budgets may have to be revised to make money available for food handouts. This diverts money from, for instance, schools and hospitals. A prolonged price slump may require handouts to farmers to keep them on the land – money that has an opportunity cost. It could have been spent on social and economic programmes for the good of the country.

5.5.2 Increased government expenditure

Some countries have put in place programmes to alleviate the condition of poor citizens. One way of alleviating their plight is to grant them transfers of money. There is a risk, however, that poor people will not use the money prudently; for instance, they may spend it on gambling or alcohol. To avoid this risk, some governments do not give money to poor people; instead, they pay for their essential items, including food. If this is the case, a rise in the price of food will increase the cost of the programme.

The USA has run a food stamp programme since the 1930s. Its objective is to alleviate the poverty of its poor citizens by, in effect, giving them free food. The programme provides some 40 million citizens (12% of the population) with food, and costs around $80 billion per year (United States Department of Agriculture, 2019).

If there is a rise in the cost of food, there is a concomitant rise in the cost of such programmes. This means that the government has to raise more revenue (more borrowing, more taxation) or cut other public programmes. The budgetary effect of a rise in the price of food is, by no means, negligible. To take the example of the USA food stamp programme, if the price of food were to rise by just 5%, this would require the government to find another $4 billion per annum (5% of $80 billion) or to cut existing programmes (see also Section 6.5 on mitigation measures).

5.6 Adverse Effects on Broader Society

5.6.1 Psychological discomfort

Having described the harm that volatile agricultural prices can bring to consumers, farmers and governments, we now consider the effects on society as a whole. Unstable grain prices bring psychological discomfort. When food prices are stable, ordinary people can go about their daily business in the knowledge that they will be able to feed their families tomorrow. Stable and predictable prices for a commodity as basic as food release families from the tyranny of now and from the exigencies of the moment. Parents can concentrate on raising and schooling their children rather than searching desperately for food to keep their children fed.

5.6.2 Difficulties in forward planning

Unstable grain prices mean that it is difficult for businesses to plan. This is particularly true for farm businesses. Should a farmer replace his 15-year-old combine harvester with a new one? If the farmer is in North America, he will have to pay $200,000 for a new model. Unless the farmer can be sure that cereal farming will be profitable in the years ahead, he may hesitate to spend so much on a single piece of machinery. If farmers are not confident of future profits, they may scale down their operations and reduce their exposure to risk. This means they produce less. Supply to the market may fall, pushing up grain prices for society as a whole. In contrast, when grain prices are stable, farm businesses can plan ahead. They can invest with greater confidence for the future and the many challenges – not least that of climate change – that the future may bring.

5.6.3 Sparking inflation

Volatile food prices can spark inflation. When citizens have to pay higher food prices, the cost of living rises. This can trigger a demand for higher wages and increased salaries. If workers are successful in their demand for more money, this can ignite an inflationary wage/price spiral.

5.6.4 The risk of unemployment

When consumers have to pay higher prices for their food, they spend less money on non-essential items. Workers in other sectors find that the demand for their goods and services begins to fall. They sell less and their incomes start to decline. Not only do they have to pay more for their food, but they also have less money coming in as wages. There is a risk of recession and unemployment.

5.7 Conclusion

Unstable prices of grain can bring big problems to consumers and farmers. Consumers can be rendered hungry and can even starve to death. Farmers can be rendered bankrupt. The work of government is disrupted. Governments may have to find more money quickly to mitigate the adverse effects for consumers or farmers. Societies do not know if they will eat tomorrow; it is difficult for businesses to make long-term plans; and food price rises can spark a wage/price spiral. Finally, if consumers have to spend more money on food it means that they have less money to spend on other goods, the demand for which may fall, putting the producers of these other goods out of a job.

Peter Timmer of Harvard University has observed that the discussion on how best to stabilize prices of commodities came to a halt in the 1980s (Timmer and Jayne, 2011). But the need to stabilize commodity prices remains as urgent as ever. For instance, Niek Koning of Wageningen University in the Netherlands states that 'international agricultural markets exhibit large price fluctuations...this volatility should be reduced' (Koning, 2017). The rest of this book examines how governments can bring about more stable prices.

Notes

[1] This, for instance, was the case with the farmers of Wollo and the pastoralists of Harerghe in Ethiopia during the droughts of the 1970s, analysed by Amartya Sen. They suffered starvation because, in his analysis, they experienced a 'direct entitlement failure.' (See Sen, 1981)

[2] In a simple example such as this, it is possible to measure the consumer product (the loaf of bread) in the same units as the raw material produced by the farmer and from which the loaf of bread has been made (the wheat). Both can be measured in 'units of wheat' or 'units of bread'. Measuring the inputs and outputs of a marketing chain in the same units is not, however, possible when there are several different inputs – for example, when wheat, sugar and cream are used to make cakes.

[3] Hansard (1963). During the course of the second reading of the Agriculture and Horticulture Bill, the UK Minister of Agriculture, Mr Fred Peart, made a speech in the House of Commons. It included the following paragraph: 'To me, it is a tragedy that as we talk about food production in this country, and about the restriction of output, in many parts of the world people are suffering from poverty and starvation. This has been said many times, but I make no apology for repeating it today. We recall the tragedies of the 1920s and 1930s, the burning of crops and the payments to farmers to induce them not to grow too much. It may be summed up in the old Canadian jingle: "Here lies the body of Farmer Pete, Who died through growing too much wheat". We do not want that to happen again. We must take the lead in creating proper world conditions. I believe that our agricultural system and our techniques are the best in the world and certainly

better than anything in Europe. Here we could give the world leadership which is essential and important. Today the tractor and the plough are much more important than the machine gun and the gunboat.'

References

Hansard (1963) Quoted from a speech in the House of Commons, by the then Minister for Agriculture, Mr Peart, 12 December 1963. Available at: https://api.parliament.uk/historic-hansard/commons/1963/dec/12/order-for-second-reading-read#S5CV0686P0_19631212_HOC_475 (accessed 11 November 2019).

Koning, N. (2017) *Food Security, Agricultural Policies and Economic Growth: Long-term Dynamics in the Past, Present and Future.* Earthscan from Routledge, Abingdon, Oxon, UK.

Neely, S. (2008) *A Concise History of the French Revolution.* Rowman and Littlefield, Lanham, Maryland.

Sen, A. (1981) *Poverty and Famines: An Essay on Entitlement and Deprivation.* Oxford University Press, United Kingdom.

Steinbeck, J. (1939) *The Grapes of Wrath.* The Viking Press, United States.

Timmer, C.P. and Jayne, T. (2011) Managing food price volatility: approaches at the global, National and household levels. Discussion at Stanford University, United States. 26 May 2011. Available at: https://www.youtube.com/watch?v=aHP9e7F11bA

United States Department of Agriculture (2019) Financial year 2019 budget summary. Available at: https://www.agri-pulse.com/ext/resources/pdfs/fy19bud-sum.pdf

6 Stabilizing Grain Prices in a Closed Economy

6.1 Introduction

Clearly, there is a need to bring a degree of stability to the grain market. This responsibility is best assumed by governments rather than by individual consumers and farmers.[1] But is it possible to stabilize grain, and if so, how? This chapter explores how to stabilize grain prices in a closed economy.

The chapter clarifies a number of terms regarding prices (6.2). It suggests that the government sets a price band. The government should try to keep the market price within this price band (6.3). It examines the economics of a buffer stock (6.4) and discusses ways of mitigating the harm done by volatile prices (6.5). The feasibility of fixing prices is considered (6.6).

6.2 Key Price Terms

First, there is the *domestic equilibrium price.* The salience of this price cannot be overemphasized. It is the key price, the lodestar on which all attempts to stabilize the price for farmers and consumers have to be focused. As Chapter 4 explained, it is the price at which supply and demand are in balance. At the domestic equilibrium price, there are no surpluses of grain left hanging over the market and there are no deficits in the form of people wishing to buy grain but finding no supplies. At this price, and only at this price, the market clears.

Secondly, there is the *fixed price.* This is the price within a controlled market, decided by the government.

Thirdly, there is the *market price.* This is the price at which people – consumers, farmers, food manufacturers, traders – buy and sell grain. It is the price in the real world, the price that is experienced whenever grain is transacted. If farmers complain that 'the price for wheat has hit a rock-bottom low', they are referring to the market price. When a man laments that 'the price of a bag of rice for my family is twice what it was a week ago', he is talking about the market price.

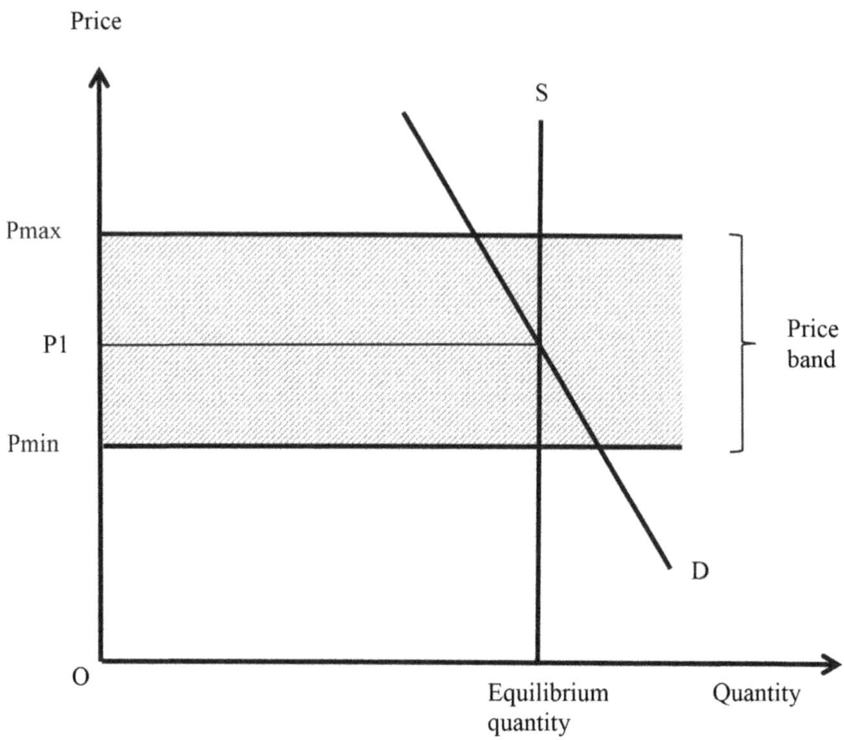

Fig. 6.1. The price band.

6.3 The Price Band

We suggest that the government's objective should be to keep the market price of grain within a 'price band'. As a first step, the government needs to set a price band. So long as the market price of grain remains within the price band, we can consider the price to be stable. This is our definition of price stability. The top of the band is the maximum price that the government considers consumers can reasonably be expected to pay for grain, given their level of disposable incomes and their other consumption needs. In the rest of this book, this price is termed the 'maximum price', abbreviated to Pmax. The bottom of the band is the minimum price which, in the estimation of the government, returns an acceptable profit to farmers for their work. In the rest of this book, it is termed the 'minimum price', abbreviated to Pmin.

A price band is illustrated in Fig. 6.1. In this particular example, the market price (P1) is equal to the domestic equilibrium price. The market price falls fairly and squarely within the price band.

If governments can keep the market price of grain within the band, neither consumers nor farmers have any ostensible cause for complaint.

The government is not faced with angry consumers or protesting farmers. Maximum and minimum prices are not a new idea by any means. Many countries have price bands for grain and for other essential commodities. For instance, the USA has specified a minimum grain price for farmers for almost 100 years, since the first Farm Bill of the 1930s.[2, 3] The European Union (EU) has, similarly, specified a minimum price for its farmers since the 1960s.[4] Although the EU does not specify a maximum price of food for consumers, whenever there is a likelihood of food prices becoming too high it takes measures to avert a rise in the price of food.

Governments need to regularly review the price band and, if necessary, modify it in line with market developments. For instance, if the economy of the country is growing at a healthy pace and people are becoming richer, they can afford to pay more for their food. In this case, the government could increase the level of the maximum price. Similarly, if farmers are adopting new technology, which enables them to reduce their production costs, then the government could be justified in lowering the minimum price.

Moreover, if for whatever reason, the long-term trend in demand (a shift in the demand curve) does not match the long-term trend in supply (a shift in the supply curve), then the domestic equilibrium price will trend up or down. The maximum price and the minimum price need to move in line with the long-term domestic equilibrium price. If the price band does not move in parallel with the long-term domestic equilibrium price, the government will find that its stabilization efforts come to nought. The government has to keep itself aligned with market trends rather than going against them and attempting to 'buck the market'.

6.4 Keeping the Equilibrium Price within the Price Band

6.4.1 The principle of a stable equilibrium price

A free market is characterized by a large number of sellers and a large number of buyers who are in open competition with each other. The price at which grain is bought and sold is the domestic equilibrium price. This is determined, as we have explained in Chapter 4, by the forces of supply and demand. Shifts in the demand curve are usually smaller and more gradual than shifts in the supply curve. The supply of grain is, to all intents and purposes, the quantity that the farmers have harvested. The size of the harvest depends on how much land the farmers have sown, and the weather. The weather is unpredictable, hence the size of the harvest is unpredictable. From one year to the next, the supply of grain can contract or expand because of the vagaries of the weather. In the context of price volatility, shifts in supply tend to be more significant than shifts in demand.[5, 6] The following discussion therefore focuses on shifts in supply of grain.

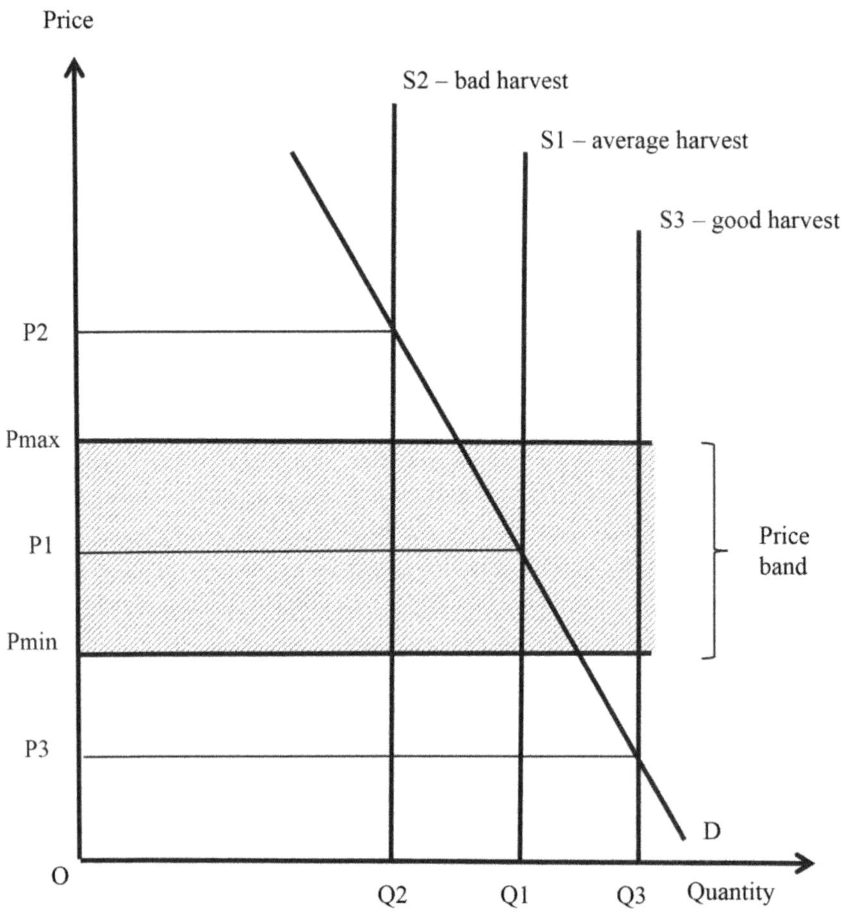

Fig. 6.2. The price of grain deviates from the price band due to shifts in supply.

If supply contracts by an excessive amount, the domestic equilibrium market price will rise above the maximum price. If supply expands by an excessive amount, the domestic equilibrium market price will fall below the minimum price. Figure 6.2 is a statement of the problem of excessive shifts in supply.

In Fig. 6.2, demand for grain remains constant at D. In a year of average harvest, supply of grain is S1. The equilibrium market price is P1. This is within the price band and there is, consequently, no price problem. Let us suppose that the harvest is below-average. Supply shifts to the left to S2. The price will rise to P2, which is above the price band. Grain will be too expensive for consumers. Now let us suppose the opposite, that the harvest is better than average. Supply shifts to the right to S3. This causes the price to fall to P3, which is less than the minimum price. This is too low for farmers to make a profit. This figure illustrates how the price can deviate from the price band depending on supply; in other words, on the size of the grain harvest. The figure states the challenge that price instability poses for governments all over the world. Governments need to find a way of countering these shifts in supply. A grain buffer stock is a way of doing this, as we now explain.

6.4.2 Expand supply by selling grain from a buffer stock

Let us suppose that the harvest is bad. Supply contracts from S1 to S2. In that case, the domestic equilibrium price will rise to P2. Food will be too expensive for consumers. The government can prevent this. It can prevent it by having a buffer stock of grain and making a commitment to sell grain, at the price of Pmax, from this buffer stock to anybody and everybody who wishes to buy it at this price. In effect, the government undertakes to become a source of supply of grain to the market at a price of Pmax and, as a result, brings about an expansion of supply. So long as the government is able to honour this commitment – that is, so long as there remains grain in the buffer stock which can be sold to consumers – the price of grain within the country will not rise above Pmax. Grain will not become too expensive for consumers. The sale of grain from a buffer stock is depicted in Fig. 6.3. Further details on the sale, and purchase, of grain are given in Chapter 9.[7]

6.4.3 Expand demand by purchasing grain to replenish a buffer stock

The above section considered a harvest that was smaller than average. Let us now consider the converse – a harvest that is bigger than average. This is illustrated in Fig. 6.4.

In Fig. 6.4, a good harvest causes supply to expand from S1 to S2. Without counter-measures, the price would fall to P2. To prevent this, the government buys grain at the price of Pmin from the country's farmers. It places the grain in its buffer stock and thus builds up the size of the stock (replenishment of the buffer stock). The government continues to automatically buy grain so long as there is downward pressure on the market price to fall below Pmin. As soon as there is no further downward pressure on the price, the government stops buying in grain for its buffer stock. Thus the signals to start buying grain and to stop buying grain come from the market. These signals trigger a government reaction (the purchase of grain) in response to the conditions in the market. The quantity that the government buys is Q2 less Q3.[8]

6.4.4 The economics of a buffer stock

A buffer stock provides a means for a government to counteract good and bad harvests. It is a physical reserve of grain that is owned by the

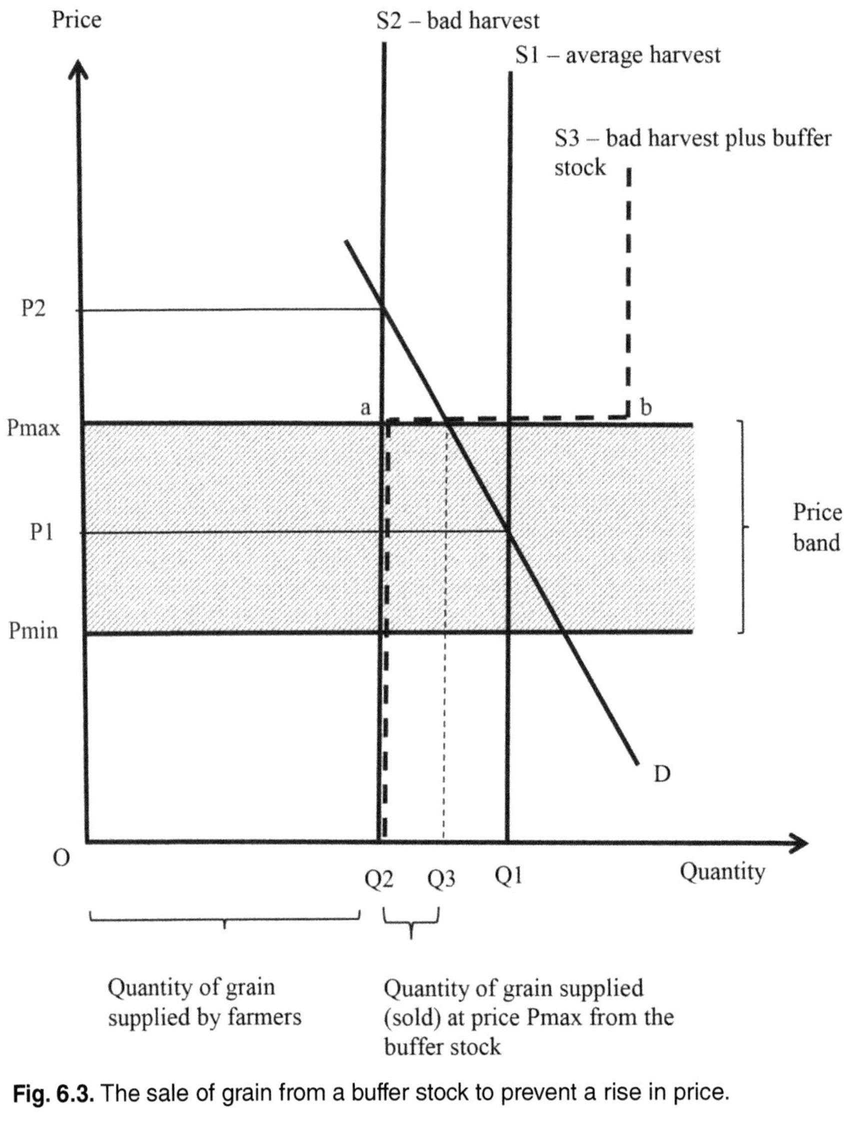

Fig. 6.3. The sale of grain from a buffer stock to prevent a rise in price.

government. The government makes two commitments: (i) to sell grain for sale on the open market at a price of Pmax; and (ii) to purchase grain from the open market whenever the market price falls to Pmin. The effect of these two commitments is to alter the shape of the supply and demand curves.

The supply and demand curves are no longer the straight curves that we have considered up to this point. By making a commitment to sell grain at Pmax, the government introduces a 'kink' into the supply curve. The supply curve becomes horizontal at Pmax, signifying that at Pmax the supply

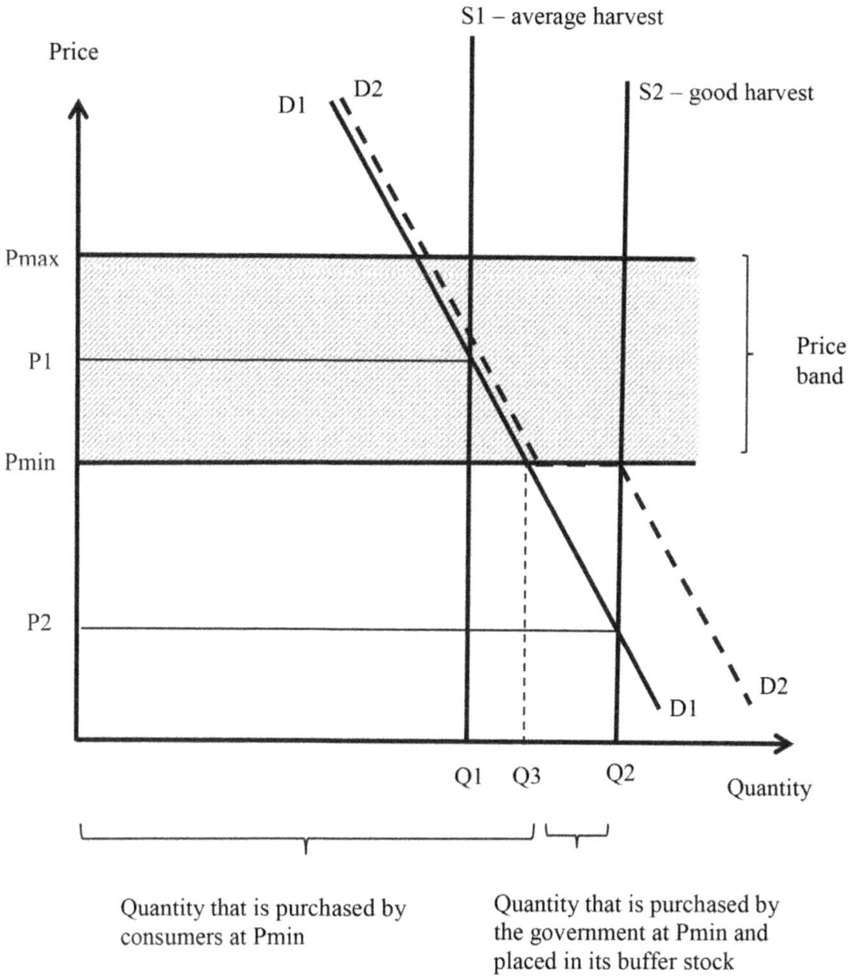

Fig. 6.4. The purchase of grain to prevent a fall in price.

from farmers is augmented by the supply from the buffer stock. At Pmax the supply becomes perfectly elastic and remains perfectly elastic until all the grain in the buffer stock is sold. The length of the horizontal section of the supply curve (the section ab in Fig. 6.3) is equal to the quantity of grain that is in the buffer stock and which the government can sell. Once all the grain in the buffer stock is sold, the government is no longer in a position to supply any further grain at Pmax (or indeed at any other price). The supply curve then becomes perfectly inelastic (i.e. a vertical line) because the only supply onto the market is from farmers. As we have noted already, the supply of grain within the season from farmers is perfectly inelastic (the curve is a vertical line) indicating that farmers are willing to sell the quantity of grain that they have harvested, irrespective of the market price.

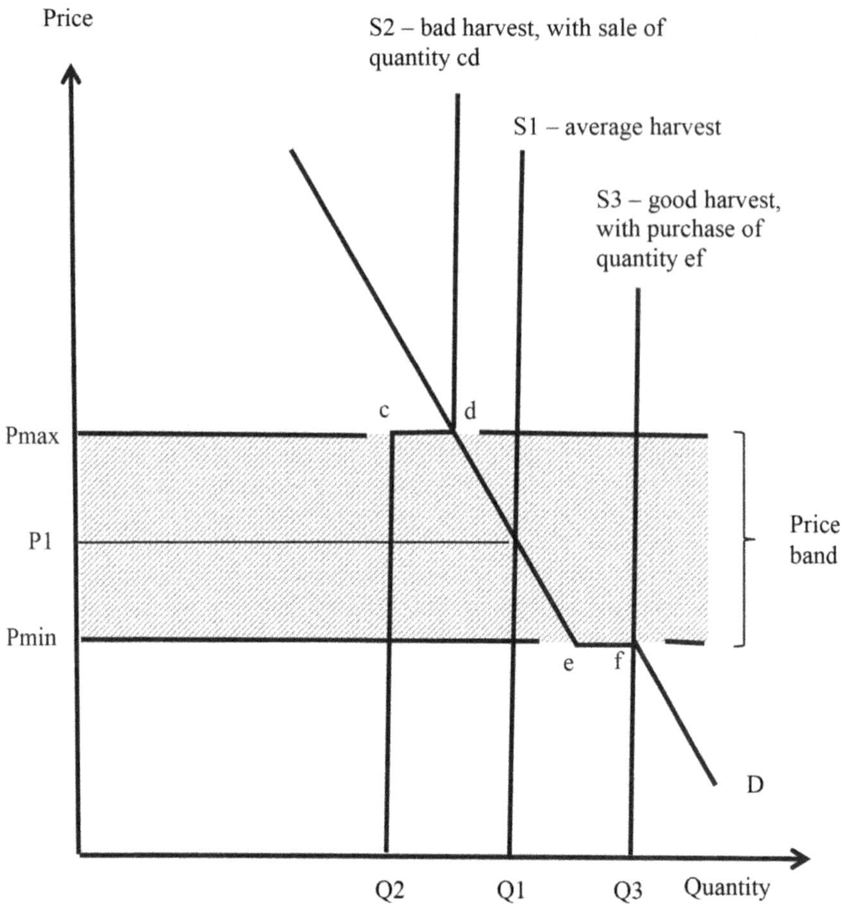

Fig. 6.5. A buffer stock keeps the price within the band despite good and bad harvests.

A buffer stock also changes the shape of the demand curve. By buying grain at Pmin, the demand curve becomes horizontal at Pmin. The commitment to purchase grain from the country's farmers is not open-ended. The government purchases only that quantity which is necessary to prevent the market price from falling below Pmin. Thus the horizontal section of the demand curve is finite – its length is equal to the quantity of grain that is purchased. At prices below Pmin and above Pmin, the demand for grain remains inelastic.

We can now recognize the significance of these 'kinks' in the supply and demand curves. Let us compare Figs 6.2 and 6.5. In Fig. 6.2, in which there is no buffer stock, we see how a bad harvest causes the domestic equilibrium price to rise above Pmax and a good harvest causes the domestic equilibrium price to fall below Pmin. Figure 6.5 shows the same fluctuations in the size of the harvest, but with a buffer stock in place. A bad harvest does not cause the domestic equilibrium price to rise above Pmax, because the reduction in the harvest is offset by the sale of grain from the buffer stock. Similarly, a good harvest no longer causes the domestic equilibrium price to fall below Pmin because the harvest 'surplus' is purchased by the government and placed in its buffer stock.

Figure 6.5 shows how a buffer stock has changed the shape of the supply and demand curves and, by so doing, keeps the equilibrium market price within the price band in years of both good and bad harvest. In a year of average harvest, the equilibrium market price is P1. In a year of bad harvest, the quantity placed by farmers on the market (Q2) is augmented by a quantity of cd sold from the buffer stock. The domestic equilibrium price is Pmax – a price that consumers can afford. In a year of good harvest, the quantity placed on the market by farmers is Q3. This would normally result in a domestic equilibrium price below Pmin. But the government expands demand at Pmin by purchasing grain for its buffer stock. It purchases a quantity equal to ef at the price of Pmin. Supply and demand are in balance at Pmin, which is therefore the equilibrium price. Farmers receive an acceptable return for their work.

6.5 Mitigating the Harm of Price Volatility

6.5.1 The principle of mitigation

A buffer stock permits the government to influence supply and demand for grain and thus to keep the domestic equilibrium price within the price band. But, inevitably, a buffer stock has limitations. First, if it is empty nobody can buy any grain from it. This is known as a 'stockout'. If there is a stockout, the government cannot prevent the market price rising above Pmax because it has no grain to sell onto the market. The opposite of a stockout can also occur and constitutes the second limitation of a buffer stock. Let us say that the national legislation specifies that the maximum size of the buffer stock is 10,000 tonnes of grain (represented by the section ab of the supply curve in Fig. 6.3).[9] In the event that there is already 10,000 tonnes of grain in the buffer stock, the government will not be able to buy more grain from its farmers. It will not be in a position to prevent the market price falling below Pmin by purchasing grain because the buffer stock is already at its maximum size. Clearly, the larger the size of a buffer stock, the less often it will run up against these limitations. Stockouts will be less frequent and there will be fewer years in which the buffer stock has reached its maximum legal size.

As an alternative to setting up a buffer stock – or as a supplementary strategy – governments can mitigate the damage to consumers of high prices and to farmers of low prices. It should, of course, be noted that although mitigation measures may alleviate the plight of consumers and farmers, the pernicious effects of price volatility on society will persist and continue to cause harm. In this section we discuss the feasibility – or otherwise – of three mitigation measures: consumer food subsidies, food rationing and deficiency payments to farmers.

Some readers may ask if food stamps (as used by, for example, the USA) and food banks (as used by the UK and other countries) mitigate the plight of consumers faced with unaffordable prices for grain. It is certainly the case that food stamps and food banks[10] provide free food to millions of families around the world. However, these programmes have become permanent income subsidies with a control of what the beneficiary can buy with the money that they receive, or in the case of food banks, with the subsidy received in kind. They are not a way of mitigating the effects of price volatility.[11]

6.5.2 Consumer food subsidies

Figure 6.6 shows the economics of a consumer subsidy applied to grain in a closed economy.

In Fig. 6.6, supply is S1, demand is D1 and the domestic equilibrium price is P1. This is within the price band and there is consequently no price problem. Let us say, however, that the harvest is bad. Supply contracts from S1 to S2. The price now escalates to P2. This is above Pmax. There is a problem because some consumers cannot afford to buy enough grain at Pmax. Let us suppose the government decides to provide its consumers with grain subsidies. This will generate a new demand curve, D2.

Why does a consumer grain subsidy generate a new demand curve? To explain this, let us first consider the case of a subsidy on a foodstuff that is continuously supplied on the market during the year, such as milk. The supply of such a foodstuff is elastic with respect to price. This is because the milk farmer can increase or decrease his production of milk from one day to the next if the price that he receives goes up or down – something that the grain farmer is not in a position to do since his is a once-a-year harvest.

Let us now suppose that the government decides to grant a subsidy of 10 cents a litre to the consumer. For every litre of milk that the consumer buys from the seller at the market price, the consumer receives 10 cents from the government. This changes the reasoning of the consumer. He, or she, now reasons that milk is cheaper by the amount of 10 cents per litre over the whole range of prices. The consumer reasons as follows:

> Before the subsidy, when the price was 20 cents per litre, I used to buy 4 litres. Now, the price that the seller asks is still 20 cents per litre, but the government gives me 10 cents per litre. So, for me, the price is really only 10 cents. Milk is now cheaper. Because it is cheaper I want to buy more of it. When the price is 10 cents, I want to buy 5 litres.

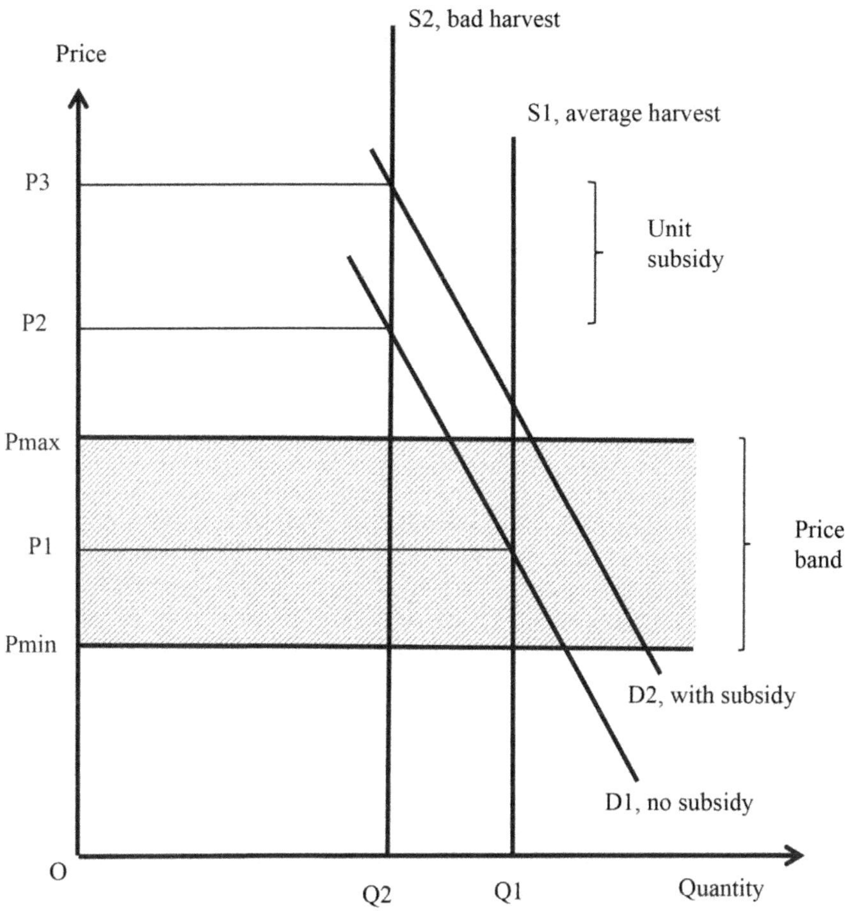

Fig. 6.6. The economics of a food subsidy (supply is perfectly inelastic).

This means that when the price in the shop is 20 cents, the consumer now buys 5 litres. Before the subsidy, the consumer used to buy only 4 litres. The consumer reasons in this manner for all the various prices in the range. This produces a new demand schedule, to the right of the original demand schedule, prior to the introduction of the subsidy. The quantity of milk on the market increases because the greater demand for milk induces a supply response from farmers.

Let us return to the question of a grain subsidy. In the case of the annual harvest of grain, the quantity of grain that is sold is fixed by the size of the annual harvest. There can be no supply response from farmers, at least not until the following year. Short-run supply of grain in a closed economy is, to all intents and purposes, perfectly inelastic and cannot be increased in response to any increase in price until the following year.

The problem with a consumer subsidy in a situation when supply is perfectly inelastic – such as grain – is that the domestic equilibrium price will rise from P2 to P3. This is because demand expands but supply is fixed. The market is seeking a new point of equilibrium in response to the greater amount of money that is now chasing grain. Quantity cannot increase, but price can increase.

Consumers will pay P3 for their grain when they buy it in the market. They will receive a subsidy of P3 less P2, meaning that the cost of food to the consumer is still P2. This is above Pmax. Thus the subsidy has not resolved the problem of unaffordable food. The beneficiaries are the farmers: before the subsidy they sold quantity Q2 at a unit price of P2. After the subsidy they sell the same quantity of grain for a unit price of P3. The point is that when the supply of grain is perfectly inelastic, food subsidies do not mitigate the harm of high food prices for consumers. Consumer food subsidies are not a plausible mitigation measure for dealing with short-run price volatility.

In economic terms, it is said that food subsidies are 'captured' by farmers. This means that the money spent by the government on food subsidies, rather than going to the consumer in the form of cheaper food, ends up in the pockets of the farmers in the form of higher prices for their grain. The price of food for the consumer stays the same. It is the farmers who are the beneficiaries.

6.5.3 Food rationing

Let us say that the harvest has been particularly poor and that the quantity of grain that is available to the country is significantly less than normal. The fact that the country is a closed economy precludes the import of supplementary grain. The poor harvest has caused grain prices to rise above Pmax. The government is forced to take measures to alleviate what can be an emergency situation with some people perhaps facing starvation. We have seen above that food subsidies do not work. What about food rationing?

The purpose of food rationing is to ensure that families who are not able to pay the equilibrium price of food (and therefore would eat less food than necessary) obtain enough food for a healthy diet. There are many ways of organizing the rationing of food. For instance, the government may distribute food for free or the government may sell it at a price that people can afford (i.e. at a price of Pmax). As for who should receive food, the government may target only those families who cannot pay a price higher than Pmax or it may provide rations to the entire population. The advantage of restricting rations to those families who cannot pay more than Pmax is that this is a smaller number of people than the entire population – therefore the cost to the government is less. However, in some circumstances it may be easier – from a political, practical and logistical point of view – to provide the entire population with rations, even though it will cost more. A third aspect is that the government may choose to set up its own food outlets to which people have to come to receive their rations or it may use the existing network of private shops, kiosks and stores.

Whatever variant the government chooses to pursue, the rationing of food as a response to a poor harvest and high prices can be an expensive and complicated matter. A main cost is that the government has to acquire the grain that it wishes to distribute. It has to buy the grain from the market. If it gives this grain away for free, it does not recoup any of the money that it has spent. It may endeavour to sell the food to the beneficiaries at a price that they can afford to pay (at Pmax) and thus recoup some of its costs. But if the situation has deteriorated into an emergency, this is ambitious, requiring accounting systems to be set up and procedures for handling cash.

A common feature of rationing schemes is the emergence of a black market. Some families may want to consume more than their allocation and be willing to pay more than Pmax for extra food. To prevent a black market, the government needs tough enforcement measures. However, as Professor Lipsey wryly remarked as long ago as 1975: 'it is an interesting comment on the strengths of various human motives that there has never been a case documented in which effective price ceilings were not accompanied by the growth of a black market. All in all, rationing food is expensive and administratively complex' (Lipsey, 1975). An additional difficulty with rationing schemes is how to avoid leakage – that is, families falsely claiming to be poor so that they can obtain food cheaply from the government rather than paying the higher equilibrium price.

For these reasons, the rationing of food in response to a sudden and unexpected increase in food prices is often an expensive and difficult exercise. It may, of course, be the only measure that is possible if the government has no food reserves that it can place on the market to make up the shortfall in the harvest. But the rationing of food is a measure of last resort.

6.5.4 Deficiency payments to farmers[12]

Governments can mitigate the harm of low farm prices by paying farmers a deficiency payment. Figure 6.7 shows the economics of a deficiency payment.

How does a deficiency payment work? The government sets a 'target price', Pmin in Fig. 6.7. This is the price that the government considers provides farmers with a minimum acceptable return for their work and investment. The government calculates the average market price during the season. This is the price that farmers have received for their grain (P2 in the Fig. 6.7). The government pays each farmer the difference between the target price and the average market price (Pmin less P2) times the quantity of grain that the farmer has sold. In this way, the government supplements

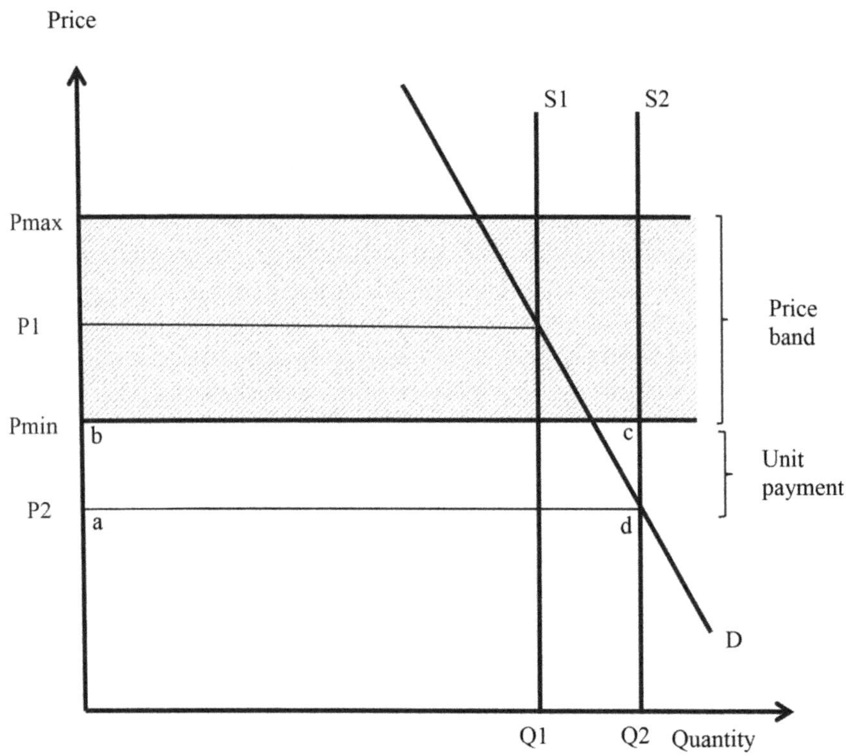

Fig. 6.7. The economics of a deficiency payment.

the farmer's income. In effect, the farmer receives the target price for each tonne of grain he sells. The government has mitigated the harm done to the farmer of a low market price.

A deficiency payment scheme has to be laid down in legislation and put into place in practice (trained staff, offices, IT equipment, auditors, accountants, rules, protocols and so forth). It can take at least a year to set up such a scheme. If a payment scheme is already set up and established, it can be an effective way of mitigating the harm done to farmers by low prices. But unless a payment scheme is already *in situ*, it may be very difficult for the government to compensate the country's many thousands, if not millions, of farmers with cash payments. The cost of deficiency payments falls on the national budget and is the multiple of the number of tonnes supplied by the farmers times the per-unit payment. This is the area abcd in Fig. 6.7.

We saw above that consumer food subsidies cannot alleviate the problem of high consumer prices due to a poor harvest, because they simply bid up the price of grain. One might expect a corresponding problem to arise with deficiency payments to farmers – that they simply push down the market price of grain. This is not the case. The quantity of grain supplied onto the market is fixed by the size of the harvest. In the short run at least, deficiency payments do not induce farmers to expand their production. Market prices are not changed by deficiency payments to farmers.

Deficiency payments tend to arouse public indignation, however. The money comes from taxpayers who are prone to ask, 'Why do farmers need to receive taxpayers money? Why cannot farmers handle price risks like all other businessmen and women?' For reasons of budgetary cost, public resistance and administrative burden, we do not consider this measure further in this book.

6.5.5 Summary of mitigation measures

To sum up this discussion on mitigation measures in a closed economy, the options are limited:

1. Food subsidies – these are not effective. The benefits of food subsidies in a closed economy are captured by farmers.
2. Rationing food – rationing schemes are expensive and administratively heavy, with the likelihood that a black market will develop.
3. Deficiency payments to farmers – these are costly and slow to set up.

6.6 Fixed Prices and Marketing Boards

The government does not have to permit a free market in its agricultural sector. Rather than allow an equilibrium price to be set for grain by the forces of supply and demand, the government may choose to set the price of grain itself. In this case there is no longer a domestic equilibrium price determined by the forces of supply and demand; there is instead a 'fixed price' determined by the government and imposed upon the market. A fixed price is fixed in the law of the land. In theory, this obviates all problems of price instability because the price can be set permanently at a level that is affordable to consumers and which also gives an acceptable return to farmers, i.e. the fixed price can be set within the price band. To ensure that the fixed price is respected and not circumvented, the government can establish a marketing board, endowed with monopsony and monopoly powers. Monopsony power means that all the farmers have to sell their produce to the marketing board – they are not allowed to sell it to anybody else. Monopoly power means that only the marketing board, and nobody else, is allowed to sell agricultural products to consumers. In other words, the marketing board stands between the farmers and the consumers. The country's entire production is channelled via the marketing board. In some circumstances, a marketing board and fixed prices may be a sensible alternative to a free market.[13] However, the main problem is that the market, as an allocative and distributive system, no longer exists. It has been supplanted by the government.

When a market is functioning well, unwritten signals in the form of equilibrium prices and equilibrium quantities are sent from all consumers to all producers and back again. The market is one large information system, provided to society at no cost. Consumers express their likes and dislikes. Producers inform consumers of what goods are on offer. This communication is two-way, informal and continuous. It ensures that producers aim to supply what consumers want at a price they can afford and that consumers buy what producers offer at a price that is remunerative.

If the government fixes the price it abolishes this communication system and replaces the market's equilibrium price with its fixed price. If the fixed price is higher that the market equilibrium price, farmers will be over-incentivized to produce. There will be a surplus of grain. If the fixed price is lower than the market equilibrium price, farmers will not have a sufficient incentive to produce the quantity that consumers wish to buy. There will be a shortage. In sum, a major problem with a fixed price is that, because it is unlikely to be set at the level of the equilibrium price, it induces the farmers to either produce too much grain or too little grain relative to the needs of the country.[14]

6.7 Summary and Conclusion

In this chapter, we have endeavoured to cover all the main options open to governments of closed economies to both prevent volatile prices and to mitigate their damage. The most attractive option is a buffer stock because it places the government in a position to strike at the root cause of the problem – harvests that are either below-average or above-average. We now turn our attention to what open economies can do to keep the price of grain within their national price band. First we need to develop a model which explains the mechanics of international trade and how the international grain market works.

Notes

[1] Could not farmers and consumers assume the responsibility for keeping prices stable, rather than the government? In reality, this is not plausible. Farmers would either have to take out an insurance policy against low crop prices or hedge the risk of low crop prices with futures contracts. Insurance firms do not offer policies against low crop prices at a reasonable premium because during a year of low crop prices every client would make a claim (Carter, 2018, p. 29).

One problem with hedging is that futures contracts do not exist for some grain crops, such as millet and teff. Where futures contracts do exist, farmers do not find them a straightforward tool to prevent price volatility. It is notable that in the USA, where farmers have been able to access futures markets for nearly 100 years, their use is very limited. A study by the US Commodity Futures Trading Commission (quoted in Carter, 2018, p. 262), found that only about 7% of US farmers use futures and that many of them were speculating rather than hedging. Professor Carter goes on to list some of the reasons given by farmers for not hedging. They include: production risk is too high and a crop shortfall could lead to futures trading losses; lack of knowledge of the hedging process; margin calls make hedging too risky; availability of forward contracts is a better alternative; and farm production does not match the size of futures contracts. As for the use of hedging by European farmers, a blog post by Madre and Devuyst (2016) explains some of the practical downsides to their use and states that only 10% of European farmers avail themselves of this approach.

As an alternative to hedging with futures contracts, farmers can hedge with options on futures contracts. Options come much closer to providing traditional price insurance than do futures because either a floor or a ceiling price can be established with options, which is not the case with futures. However, according to Carter (2018, p. 356), in many instances, options are not necessarily superior to futures as a hedging tool, largely because trading options require the buyer of the option (i.e. the farmer) to pay an upfront premium to the seller of the option. In a volatile market, the premium may be very high.

For these practical reasons, we feel that it is unrealistic to expect farmers, by themselves, to mitigate the damage caused to them by price volatility through the use of insurance, futures contracts and options on futures contracts. Furthermore, even if farmers did mitigate the damage caused to them, grain prices would still be volatile and cause damage to the society as a whole. There is, therefore, a role for the state to moderate the volatility of grain prices.

As for consumers assuming the responsibility of ensuring that they always have a sufficient supply of food (as opposed to the state assuming this responsibility), this would require that all consumers keep a food reserve. Whilst some consumers could be convinced to take such a measure, it seems unlikely that the poorer sections of society would be able to afford to do so.

[2] For instance, in the 2014 Farm Bill (Public Law 113–79) the minimal price of wheat was set at $5.50 per bushel.

[3] The USA operates a system of parity prices. Paarlberg (1964, p. 69) explains the concept as follows: If a bushel of wheat exchanged for a pair of overalls years ago, a bushel of wheat ought to exchange for a pair of overalls today. This is the basic idea of parity. The base period is 1910–1914. Production goods, wages and household items are included in the index of prices of articles bought by farmers. The arithmetic of the parity computation, stripped of complications, is as follows: the ratio of the price of wheat per bushel in 1910–1914 to the index of prices of articles bought by farmers is to equal the parity price of wheat today to the index of articles bought by farmers today. This is the formula by which the parity price for US farmers is calculated. The intention is to ensure that farming remains as profitable today as it was during the period just prior to World War I.

[4] For instance, the EU adopted a regulation regarding minimum prices (regulation 1308/2013) in which the minimum price for wheat was set at EUR 101,31/tonne. The legal term used by the EU is 'reference threshold' while in common parlance it is known as a 'safety net price'.

[5] As stated in the main text, demand for an agricultural product can change from one year to the next, but invariably, it does so in a gradual manner and does not induce a large and sudden change in price. Such changes are due to a change in the size of the population, a change in income or a change in dietary preferences.

These are changes in long-run demand. But this is not to say that sudden and unexpected shifts in the demand, i.e. short-run demand, for an agricultural product never occur. They can do so, but they are rare. They can be caused by a food scare, such as that which occurred in Europe when a disease of cattle ('mad cow disease') was discovered in the 1990s. The short-run demand for beef fell precipitously since consumers wanted to avoid eating meat that could have been contaminated with this disease.

[6] Does panic buying cause the short-run demand curve to shift to the right? Surely, if consumers rush to the shops to buy food, this constitutes an expansion of demand. If so, panic buying would be a cause of the type of price volatility that we are considering in this book. Panic buying can occur when it is known that a country's harvest will be poor and that there may be a shortage of food. Consumers may also panic if there is a rumour that the local supermarket will run out of bread, sugar or another essential food item. In reality, panic buying is unlikely to cause the short-run demand curve to shift perceptibly. The reason is that the short-run demand curve depicts the annual demand – the quantity that consumers demand throughout the course of the entire year. Panic buying is likely to accelerate the timing of purchases but it is unlikely to increase the quantity that is demanded during the year as a whole.

[7] In theory, governments can expand supply of grain by making a request for food aid from the United Nations World Food Program (headquarters in Rome). This is an alternative to selling grain from a national buffer stock and has the same effect of preventing a rise in the equilibrium price. However, the World Food Program is not adequately financed and is not a reliable or timely source of supplementary grain.

[8] Rather than purchasing grain from the market, which causes the demand curve to shift to the right, could not the government restrict the supply of grain from farmers, thereby shifting supply to the left? Would not such 'production quotas' or 'supply quotas' be an alternative way to prevent a fall in price? A production quota is a tool for bringing about a reduction in planned supply, not actual supply. This book is concerned with actual supply – the quantity of grain (or of another agricultural commodity) that farmers actually produce, not the quantity that they plan to produce when they plant their fields at the beginning of the season. If the government wishes to reduce the quantity that farmers actually place on the market (rather than plan to place on the market), a supply quota is necessary. This then begs the question: What is supposed to happen to that grain that is harvested but not placed on the market? Either it can be destroyed (which is a waste of resources and likely to arouse the consternation of the public) or it can be retained by farmers until the following season. But further questions arise: Who is going to pay for the cost of storage; How is the government supposed to stop farmers from illicitly selling the grain that is in excess of the supply quota and which is not to be marketed until the following year? For these practical reasons, supply controls are not considered in this book.

[9] Rather than specifying the maximum physical size of a buffer stock (e.g. 10,000 tonnes), the government may specify a limit on the amount of public funds that can be tied up at any one time in purchased grain (e.g. US$1 million). Rather than a physical limit there is now a budgetary limit.

[10] How much do food banks cost? Data is scarce because these schemes are privately run by charities and churches. The Trussel Trust is a food bank charity in

the UK running some 1500 food banks, with 40,000 volunteer workers. Its annual report for 2018 reported that it had given away 1.3 million three-day food parcels in 2017/18. Its income during the year was £ 7.3 million. (Source: 31 March 2018 Annual Report and Accounts, Trussel Trust, Salisbury, UK.)

[11]If food stamps programmes and food banks are already in place, and if the government increases the budgetary allocation in line with the increase in the price of grain-based foods, such programmes will shield their beneficiary families from rises in the price of food. In that case, the rise in the price of food is borne not by consumers but by the government (as mentioned in Section 6.5.2). The programmes have, in effect, removed this group of consumers from the market because their food needs are now met by the government.

[12]Deficiency payments are to be distinguished from other types of payments to farmers. Deficiency payments are designed to give farmers a guaranteed minimum price for their products. Other types of payment have other objectives – for instance, the direct payments paid to farmers under the EU's Common Agricultural Policy since the mid-1990s have the purpose of boosting farmers' incomes. They do not provide farmers with a minimum price for their products.

[13]In practice, it can be difficult to ensure that all farmers sell all their produce to a marketing board. In the case of those commodities – such as milk and sugar – which require processing, if the marketing board is the sole processor, farmers will have no choice but to sell to it. But with other products, such as feedgrains, it can be difficult to prevent parallel sales, i.e. preventing a cereal farmer directly selling his barley to a dairy farmer who uses it to feed his dairy cows. In this case, the cereal farmer bypasses the marketing board.

[14]Although not focused on the agricultural sector, Krugman and Wells, 2015, chapter 5, is a good discussion of the practical problems that can arise when governments attempt to impose fixed prices, in the form of ceiling and floor prices, on a market.

References

Carter, C. (2018) *Futures and Options Markets: An Introduction.* Rebel Text, Davis, California.

Krugman, P. and Wells, R. (2015) *Microeconomics,* 4th edn. Worth publishers, New York.

Lipsey, R. (1975) *An Introduction to Positive Economics,* 4th edn. Weidenfeld and Nicolson, London.

Madre, Y. and Devuyst, P. (2016) Are futures the future for farmers? Farm Europe, posted 29 April 2016. Available at: https://www.farm-europe.eu/travaux/are-futures-the-future-for-farmers-2/ (accessed September 2019).

Paarlberg, D. (1964) *American Farm Policy: A Case Study of Centralized Decision-making.* John Wiley, New York.

A Model of the International Grain Market

7

7.1 Introduction

The previous chapter considered how, when faced with good and bad harvests, the government of a closed economy could stabilize the market price within its price band. It concluded that a buffer stock is the most plausible option. A buffer stock places the government in a position to influence the supply of grain and the demand for grain on its national market. By influencing supply and demand, the government is able, in turn, to influence the level of the domestic equilibrium price.

Most countries, however, trade with others. They are not closed economies; they are open economies. It is to these countries that our discussion now turns. Is it possible for the government of an open economy to stabilize the price of grain? Or does an open economy simply have to accept the world price, irrespective of whether it is high or low? To help us answer these and other questions, we need a model of international trade. It is the purpose of this chapter to construct such a model.

We start by reminding ourselves of the benefits of trade (7.2). We then proceed, step-by-step, to develop a model of international trade. First, we introduce two new concepts: the supply of exports; and the demand for imports (7.3). We then imagine a world of two countries, A and B, which are trading with each other. We deploy our two new concepts to describe the trading relationship between them. The trading relationship has three elements: the direction of trade (importer or exporter?), the quantity of grain that is traded and, of great importance to our core concern of price volatility, the 'world price', being the price at which grain is traded (7.4).

Next we expand our imaginary world of two countries to a world of four countries. We show how the model permits the trading relationship of any one individual country *vis-à-vis* the rest of the world to be revealed. The rest of the world in this case consists of the remaining three countries (7.5). The model can reveal whether a particular country is an importer or exporter, how much grain it trades and the price at which it buys and sells grain. It can describe the trading relationship of any one country with the rest of the world.

The question then arises: Do individual countries affect the world price? For instance, what would be the impact if one of the four countries in our imaginary world were to suffer a poor harvest? To make up for its low domestic production of grain, the country might be obliged to import more from the rest of the world. Would this drive up the world price, or would the world price remain unaffected? Conversely, if one of the four countries was to enjoy an exceptionally good harvest, what would be the impact, if any, on the world price? Would the world price fall, due to an expansion of its supply of imports arising from the country's very good harvest?

The answers to such questions depend on whether the country is 'large' or 'small' in terms of its share of world trade. This is discussed in Section 7.6. It explains why we can assume that the size of the harvest of small countries has no influence on the world price, whereas the size of the harvest of large countries does have an influence on the world price. This leads us to the conclusion that small countries are 'price takers', while large countries are not.

7.2 The Benefits of Free Trade

The theory behind free trade was first articulated in 1817 by the British economist David Ricardo. He studied trade between nations, not in terms of absolute advantage but in terms of comparative advantage (Ricardo, 2004). When countries specialize in producing what they are relatively best at – compared to other countries – all countries are better off. Everybody's welfare can be increased. Ricardo formulated what has become known as the Law of Comparative Advantage. To this day, it provides the rationale for the pursuance of free trade and for the avoidance of trade measures – such as import duties, export quotas and so forth. Ricardo's insight into the effects of trade was received favourably by the government of the day and lay behind the reform of Britain's trade policy. It was partly because of Ricardo that Britain rescinded its tariffs against imports of foreign grain and embraced free trade. This was the first time that a government changed its policy on the basis of economic theory.

7.3 Two Concepts: The Supply of Exports and the Demand for Imports

In Chapter 4 we developed a model of the grain market in a closed economy. We now need to take that same model but modify it so that it can describe a country that is trading with the rest of the world, as most countries currently do.

Table 7.1. Country A – supply, demand, supply of exports and demand for imports (bushels).

Price ($ per bushel)	Quantity supplied	Quantity demanded	Supply of exports	Demand for imports
(1)	(2)	(3)	(4=2 less 3)	(5=3 less 2)
10	80	70	10	
9	80	75	5	
8 – domestic equilibrium price	80	80	0	0
7	80	85		5
6	80	90		10
5	80	95		15
4	80	100		20
3	80	105		25
2	80	110		30
1	80	115		35

Let us start with country A. Its supply and demand schedules are shown in Table 7.1. We see that at the price of $8 per bushel, the quantity supplied (80 bushels) and the quantity demanded (80 bushels) are in balance. This price is the domestic equilibrium price. Up to now, this book has not considered the possibility that the market price deviates from the domestic equilibrium price. When countries trade with each other, the market price is not the same as the domestic equilibrium price.

If the market price is above the domestic equilibrium price, more grain is supplied than is demanded. For instance, if the price is $9 per bushel, the quantity supplied is still 80 bushels (supply does not vary with price because the size of the harvest is fixed at 80 bushels), but the quantity demanded is 75 bushels. There are five bushels supplied by farmers but not purchased by the consumers of the country. These five bushels are exported. We say that at prices above the domestic equilibrium price, there is a 'supply of exports'. The higher the price, the more exports are supplied.

As for prices below the domestic equilibrium price, there is a 'demand for imports'. Suppose the price is $5 per bushel. The quantity supplied by farmers stays at 80 bushels but the quantity of grain demanded by consumers in country A increases to 95 bushels. The farmers do not satisfy the demand of their own consumers. So from where do the missing 15 bushels come? They come from abroad in the form of imports. The lower the price, the greater the quantity of imports that is demanded. In Table 7.1, columns 4 are 5 are the arithmetical differences between the quantity supplied and the quantity demanded, over the range of prices considered, and show the supply of exports and the demand for imports.

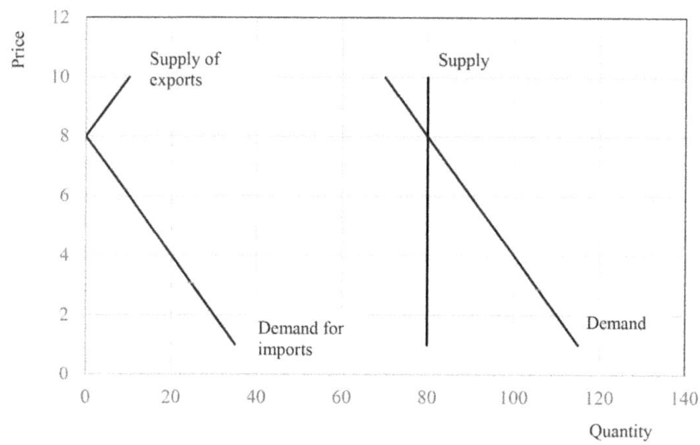

Fig. 7.1. Country A – supply of exports and demand for imports.

We can plot the information given in Table 7.1 in Fig. 7.1.

7.4 A World of Two Countries

Let us now take a second country – country B. Table 7.2 and Fig. 7.2 show its supply and demand features. It is similar to country A in that it supplies a quantity of 80 bushels. The difference is that, at any given price, its consumers demand a smaller quantity of grain.

Table 7.2. Country B – supply, demand, supply of exports and demand for imports (bushels).

Price ($ per bushel)	Quantity supplied	Quantity demanded	Supply of exports	Demand for imports
(1)	(2)	(3)	(4=2 less 3)	(5=3 less 2)
10	80	60	20	
9	80	65	15	
8	80	70	10	
7	80	75	5	
6– domestic equilibrium price	80	80	0	0
5	80	85		5
4	80	90		10
3	80	95		15
2	80	100		20
1	80	105		25

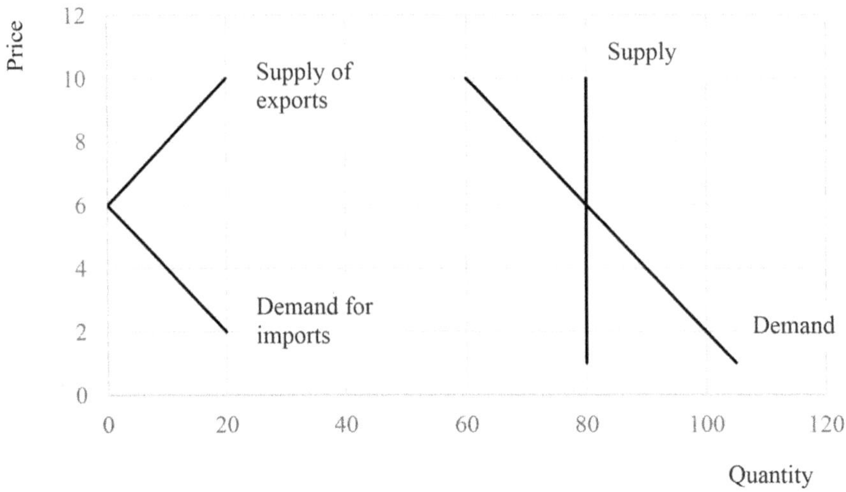

Fig. 7.2. Country B – supply of exports and demand for imports.

Suppose now that the two countries are able to trade with each other. The consumers of country A are able to demand grain from both their own farmers and from the farmers of country B. Likewise, the farmers of country A can supply their own consumers and the consumers of country B. The new situation is given in Table 7.3. In this table, the exports supplied by country A at any given price are added to the exports supplied by B at the same price. The total exports are given in column 4. Similarly, the imports demanded by country B at any given price are added to the imports

Table 7.3. Countries A and B trade with each other.

Price ($ per bushel)	Supply of exports (bushels)			Demand for imports (bushels)		
	A's exports	B's exports	Total exports	A's imports	B's imports	Total imports
(1)	(2)	(3)	(4=2 plus 3)	(5)	(6)	(7=5 plus 6)
10	10	20	30			
9	5	15	20			
8	0	10	10	0		0
7		5	5	5		5
6		0	0	10	0	10
5				15	5	20
4				20	10	30
3				25	15	40
2				30	20	50
1				35	25	60

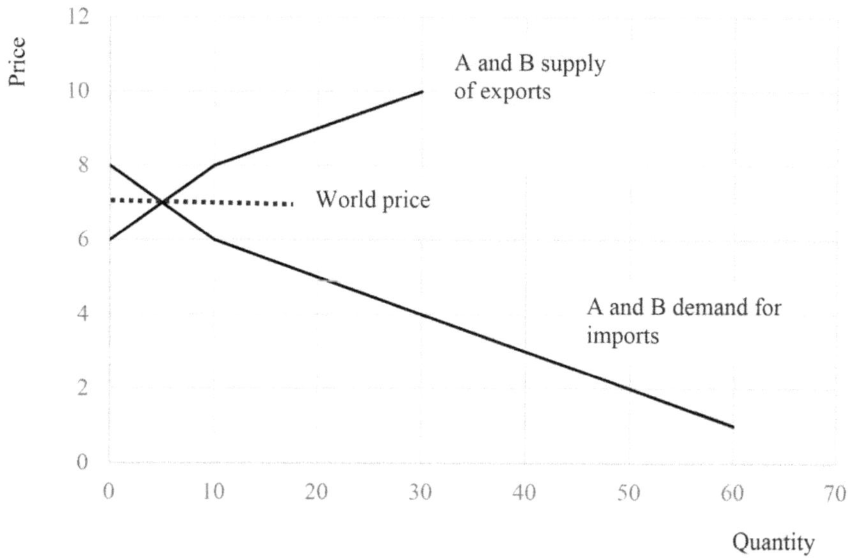

Fig. 7.3. Countries A and B trade with each other.

demanded by country B at the same price. The total imports are given in column 7.

We can portray this new situation by plotting the total supply of exports of the two countries and the total demand for imports of the two countries. This is shown in Figure 7.3.

Figure 7.3 shows two curves. One curve is the supply of exports from countries A and B when they are considered to be one and the same entity. It is column 4 of Table 7.3. The curve consists of two sections. The first section is from \$6 to \$8. It describes the supply of exports from country B only (column 3). Country A does not feature in this first section. This is because it does not supply any exports over this range of prices. It begins to supply exports only when the price has surpassed \$8 (column 2). The curve therefore has a kink at \$8 per bushel, this being the price at which the supply of exports from country B begins to be supplemented by the supply of exports from country A.

The second curve is the demand for imports into countries A and B, considered as a single entity. It is column 7 of Table 7.3. Like the curve that describes the supply of exports, this curve also consists of two sections. The first section is from \$8 to \$6. It describes the demand for imports into country A only (column 5). Country B does not express any demand for imports over this range of prices – it begins to demand imports only when the price has fallen below \$6 per bushel (column 6). At prices below \$6 per bushel, the demand for imports into country A begins to supplement the demand for imports into country B. This is shown by the curve having a kink at a price of \$6 per bushel.

The two curves intersect at a price of $7 per bushel. On closer inspection we can see that it is the first sections of both curves that intersect. It is not the second sections that intersect. The first section of the supply of exports curve describes the exports of country B. The first section of the demand for imports curve describes the imports of country A. Therefore the price of $7 per bushel is the price that exactly balances the supply of exports from country B with the demand for imports into country A.

The quantity of grain exported is five bushels (it is exported by country B) and the quantity of grain imported is also five bushels (it is imported by country A). The price of $7 per bushel is the price at which grain is exchanged between the two countries. It is also the price that pertains within the two countries. The domestic equilibrium price of $8 per bushel in country A no longer exists. Neither does the domestic equilibrium price of $6 per bushel exist in country B. A new price of $7 per bushel pertains in country A, in country B and between the two countries. It is the 'world price'.

7.5 A World of Four Countries

Let us now expand our world of two countries. Let it consist of four countries, A, B, C and D. The supply and demand characteristics of the two new countries are shown in Table 7.4, while Table 7.5 assembles the supply of exports and the demand for imports of each country, and provides the total.

Figure 7.4 shows the total supply of exports and the total demand for imports.

Like Fig. 7.3, Fig. 7.4 consists of two curves. The curve which slopes upwards describes the aggregate supply of exports, over the full range of prices, from countries A, B, C and D considered as a whole. It is column 6 of Table 7.5. The curve consists of sections, delineated by kinks. This reflects the fact that each country has its own supply-of-exports curve. The kinks occur at those prices when a country begins to supply exports of grain. The first section of this curve runs from $2 to $4 per bushel. It describes one country only – country D – since it is only country D that supplies any grain for export when the price is between $2 and $4 (column 5). The second section of the curve runs from $4 to $6. Over this price range, country D continues to export grain and it is joined by country C (column 4). The second section therefore describes countries C and D. As prices increase further – that is, as they rise above $6 per bushel and beyond – the remaining two countries begin to supply exports and new sections of the curve are accordingly added.

The second curve is the curve which slopes downwards. This describes the demand for imports into the four countries, considered as a single and whole entity. It, too, consists of sections, delineated by kinks. This reflects

Table 7.4. Countries C and D – supply of exports and demand for imports (bushels).

	Country C				Country D			
Price ($ per bushel)	Supply	Demand	Supply of exports	Demand for imports	Supply	Demand	Supply of exports	Demand for imports
10	80	50	30		80	40	40	
9	80	55	25		80	45	35	
8	80	60	20		80	50	30	
7	80	65	15		80	55	25	
6	80	70	10		80	60	20	
5	80	75	5		80	65	15	
4	80	80	0	0	80	70	10	
3	80	85		5	80	75	5	
2	80	90		10	80	80	0	0
1	80	95		15	80	85		5

Table 7.5. A world of four countries – supply of exports and demand for imports (bushels).

Price ($ per bushel)	Supply of exports					Demand for imports				
	A	B	C	D	Total	A	B	C	D	Total
(1)	(2)	(3)	(4)	(5)	(6)	(7)	(8)	(9)	(10)	(11)
10	10	20	30	40	100					
9	5	15	25	35	80					
8	0	10	20	30	60	0				0
7		5	15	25	45	5				5
6		0	10	20	30	10	0			10
5			5	15	20	15	5			20
4			0	10	10	20	10	0		30
3				5	5	25	15	5		45
2				0	0	30	20	10	0	60
1						35	25	15	5	80

the fact that each of the four countries has its own demand-for-imports curve. The first section of the curve is from $8 to $6 per bushel and describes the demand for imports into country A (column 7). The second section is from $6 to $4 per bushel and describes the demand for imports into both countries A and B. As prices further decrease, so the remaining two countries begin to demand imports. Accordingly, further sections are added to the curve.

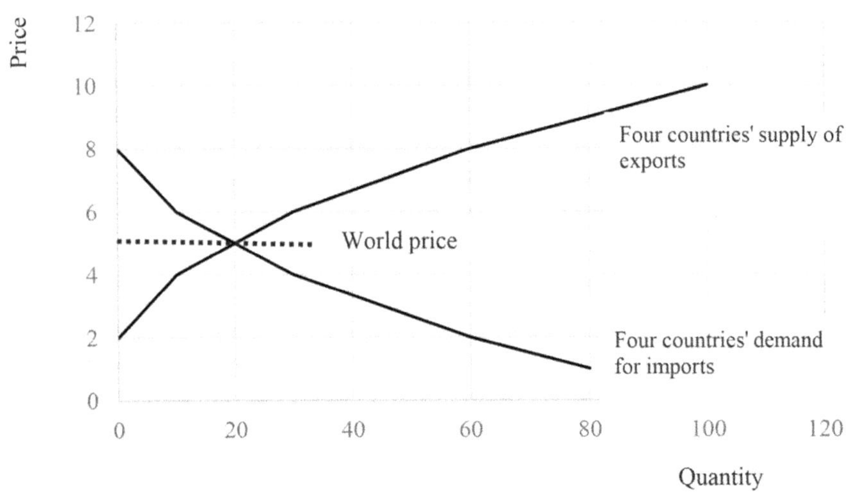

Fig. 7.4. A world of four countries – supply of exports and demand for imports.

Having described the make-up of these two curves, what does Fig. 7.4 tell us? It reveals three points. First, it tells us that when the price is at a level of $5 per bushel, the four countries find a balance of imports and exports among themselves. At this price, all those countries that want to export grain can find a buyer for it. Likewise, all those countries that want to import grain can find a seller. There is no country wanting to import but unable to find another country willing to sell to it. Similarly, there is no country that wants to export grain but is unable to find a buyer for it. The price of $5 per bushel is the price that brings balance to the international grain market. This is the world price and at this price the market clears.

Secondly, it tells us that the quantity of grain that is exported and imported is 20 bushels. This is the point on the quantity axis at which the two curves intersect. Thirdly, the figure tells us which countries are importing grain and which countries are exporting it. Looking closely at the figure, we see that it is the second section of each curve that intersects the other. The second section of the supply-of-exports curve describes the exports of countries C and D. It is, therefore, countries C and D that are the exporters of grain in our world of four countries. Looking along the row in Table 7.5, which describes the situation when the price is $5 per bushel, we can see that country C exports five bushels and country D exports 15 bushels. The second section of the demand-for-imports curve describes the imports of countries A and B. These two countries are the importers of grain in our world of four countries. Again, looking along the row in Table 7.5, which describes the situation when the price is $5 per bushel, we see that country A imports 15 bushels and country B imports five bushels.

What we have done is to construct a model of international trade between four countries. The model is valuable because it can reveal the price that will prevail both between and within countries, the quantity of grain that will be exported and imported and, finally, which countries will import and which countries will export.

7.6 A Single Country in the Context of the Rest of the World

Let us now exploit some of the potential of this model. We need to be able to explore how a single country is affected when it trades with the rest of the world. This will enable us to explore how, on the one hand, events in the rest of the world impact an individual country and, on the other hand, how these same events in the country itself may affect the rest of the world. The sorts of events that the model will help us explore are events such as good harvests, bad harvests, taxes on exports and subsidies on imports.

To modify our model, we can choose an individual country and extract it from the world of four countries that we have created. Let us extract country D. That leaves countries A, B and C as the rest of the world.

Table 7.6 shows the supply and demand schedules for country D and the rest of the world, while Fig. 7.5 shows the same information in graphical form.

Figure 7.5 describes country D in the context of the rest of the world. There are five lines in this figure. What do they all mean? The two continuous lines are the schedules for country D itself – the supply of exports (column 4 of Table 7.6) sloping up to the right and the demand for imports sloping down to the right (column 5). The domestic equilibrium price inside country D is $2 per bushel. At this price there is neither a surplus of supply nor an unsatisfied demand.

Turning now to the two dashed lines, these are the schedules for the rest of the world – that is, countries A, B and C as a whole. The supply of exports from the rest of the world is shown as the upward-sloping curve and depicts column 8, while the demand for imports into the rest of the world is shown as the downward-sloping curve, depicting column 9. The domestic equilibrium price inside the rest of the world is $6 per bushel. At this price there is no supply of exports and no demand for imports.

The two curves that are important to us are D's supply of exports and the rest of the world's demand for imports. The point of market balance is at the intersection of these two curves. This is at a price of $5 per bushel and a quantity of 15 bushels. The price of grain – within the four countries themselves and between the four countries – is $5 per bushel. Country D supplies a quantity of 15 bushels to the rest of the world.

When we compare Figs. 7.4 and 7.5, we see that both figures show that the world price is $5 per bushel. However, the quantity that is traded is not the same. Figure 7.4 shows that the traded quantity is 20 bushels, while Fig. 7.5 shows a traded quantity of 15 bushels. There is a difference of five bushels. So why have these missing bushels disappeared? This is the grain that is traded between the countries that form the rest of the world. These bushels appear in Fig. 7.4 but are not revealed in Fig. 7.5 because the latter figure shows the rest of the world as it presents itself to country D. It does not show the trade that is taking place between the countries themselves.[1]

7.7 The Impact of a Country's Harvest on the World Price

A model of this nature can help us explore a wide variety of questions regarding international trade in grain. In the current context, we are, of course, interested in the price at which grain is traded – in other words, in the world price of grain. We have seen above that this is determined by the demand for imports and the supply of exports. The world price is that price at which these two forces are brought into balance, as shown in Fig. 7.4. This being so, it follows that when either of these two forces changes, then the world price also changes. In other words, if the supply of

Table 7.6. Country D and the rest of the world (countries A, B and C) (bushels).

Price ($ per bushel)	Country D				The rest of the world			
	Supply	Demand	Supply of exports	Demand for imports	Supply	Demand	Supply of exports	Demand for imports
(1)	(2)	(3)	(4)	(5)	(6)	(7)	(8)	(9)
10	80	40	40		240	180	60	
9	80	45	35		240	195	45	
8	80	50	30		240	210	30	
7	80	55	25		240	225	15	
6	80	60	20		240	240	0	0
5	80	65	15		240	255		15
4	80	70	10		240	270		30
3	80	75	5		240	285		45
2	80	80	0	0	240	300		60
1	80	85		5	240	315		75

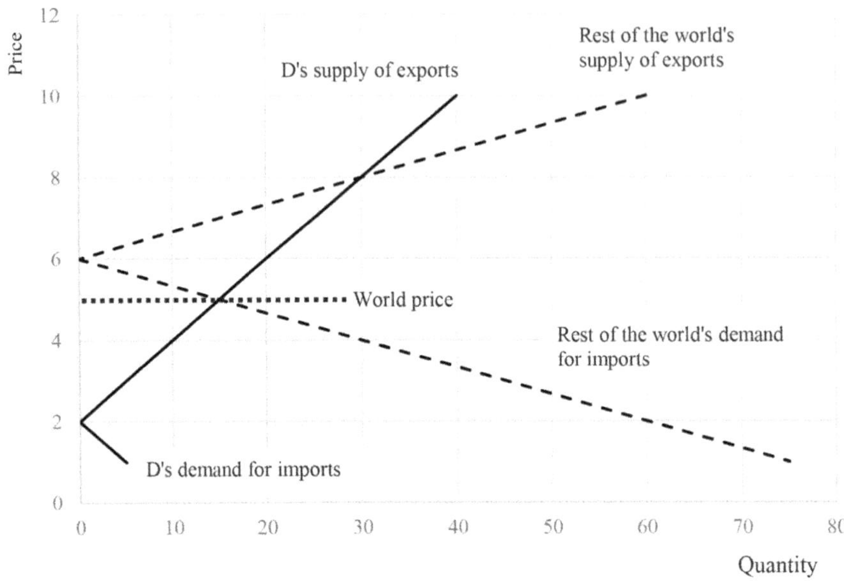

Fig. 7.5. International trade between country D and the rest of the world (countries A, B and C).

exports either contracts (shifts to the left) or expands (shifts to the right), this will bring about a change in the world price. Similarly, if the demand for imports either contracts (shifts to the left) or expands (shifts to the right), this will, in its turn, bring about a change in the world price.

A country's demand for imports and supply of exports is the difference between the quantity supplied by its farmers (the harvest) and the quantity demanded by its consumers over the range of prices considered. We have supposed above, in Table 7.6, that the harvest of country D is 80 bushels. Let us consider this to be the average harvest. In some years the harvest is better; it is 90 bushels. This constitutes a good harvest. But in other years the harvest is bad. Farmers reap only 70 bushels. Let us assume that demand for grain within country D remains the same. Table 7.7 shows the situation in years of good and bad harvests.[2] This information is plotted in Fig. 7.6 (good harvest) and Fig. 7.7 (bad harvest).

Figure 7.6 shows the effect of a good harvest in country D on, first, the country's supply of exports and, secondly, on the world price. The effect of a good harvest is to expand the country's supply of exports – this is depicted by the shift to the right in the supply of exports. As a result the supply of exports now intersects the rest of the world's demand for imports at a lower price. In a year of average harvest, the world price is $5 per bushel. But when country D enjoys a good harvest, the world price falls to $4.5 per bushel.

Table 7.7. Country D – good and bad harvests (bushels).

	Good harvest				Bad harvest			
Price ($ per bushel)	Supply	Demand	Supply of exports	Demand for imports	Supply	Demand	Supply of exports	Demand for imports
(1)	(2)	(3)	(4)	(5)	(6)	(7)	(8)	(9)
10	90	40	50		70	40	30	
9	90	45	45		70	45	25	
8	90	50	40		70	50	20	
7	90	55	35		70	55	15	
6	90	60	30		70	60	10	
5	90	65	25		70	65	5	
4	90	70	20		70	70	0	0
3	90	75	15		70	75		5
2	90	80	10		70	80		10
1	90	85	5		70	85		15

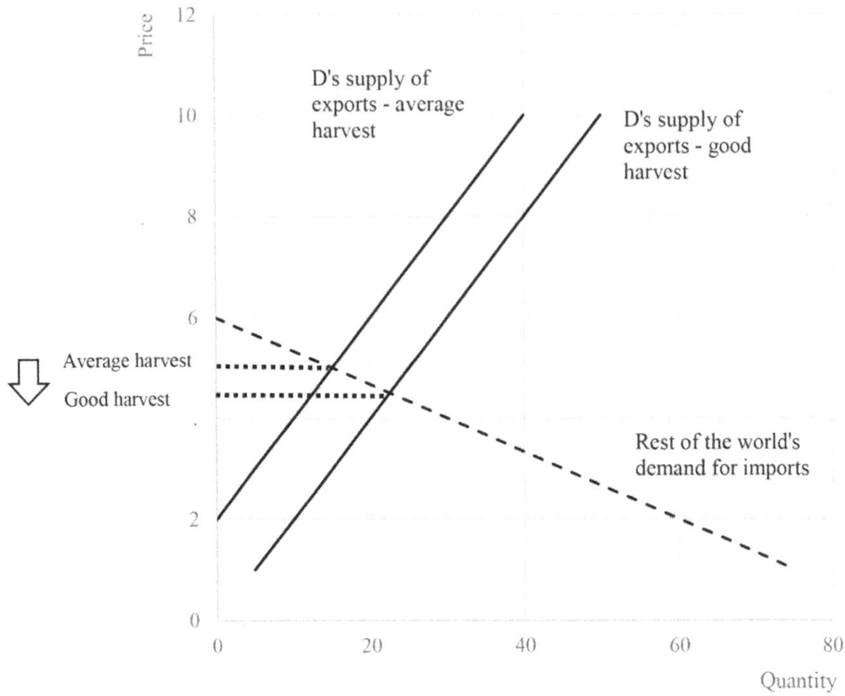

Fig. 7.6. Country D – the effect of a good harvest on the world price.

In contrast, Fig. 7.7 shows that when the harvest of country D is bad, there is a rise in the world price.

7.8 'Large' and 'Small' Countries

We have been considering an imaginary world of four countries. In this world, D is a large player. In an average year, its domestic supply (the harvest) is 80 bushels. The other three countries, similarly, each supply 80 bushels. Thus country D supplies 25% of the total world supply. This makes it a 'large' country relative to the rest of the world.

In reality, very few countries are large enough for changes in their harvest to have a discernible effect on the world price. Depending on the magnitude of any change in their supply of exports or their demand for imports, countries such as China, USA, Russia, Ukraine, Canada and the EU may have an impact on the world price. Many other countries in the world can be considered to be 'small'. The world does not consist of four countries, each supplying a quarter of world grain production. Rather, it consists of 200 countries. If all the countries were the same size in terms of grain production, each would supply only 0.5% of world grain production. This is very much less than the quarter supplied by each country in

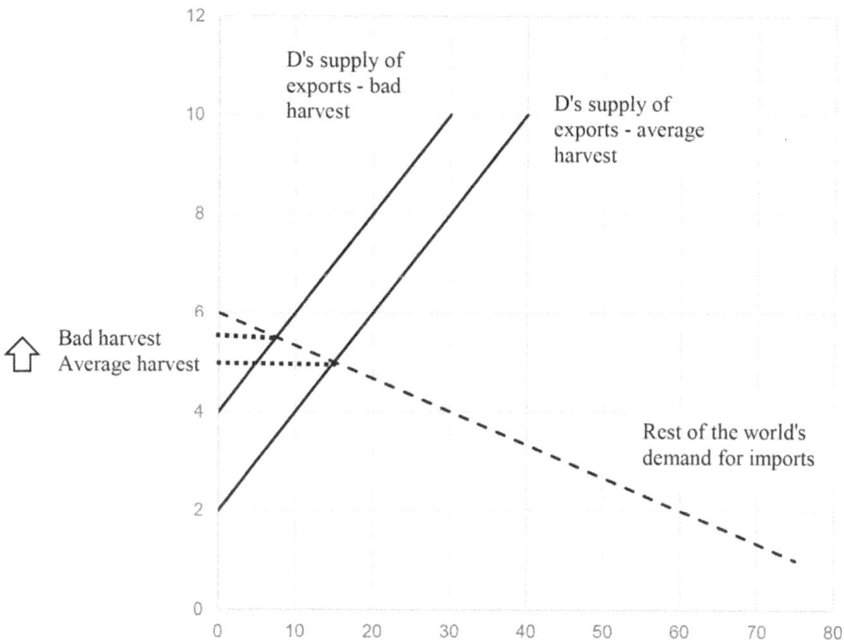

Fig. 7.7. Country D – the effect of a bad harvest on the world price.

our imaginary world. Moreover, many countries are much smaller that the average size and, as a result, supply less than 0.5% of world production. Relative to the rest of the world, an individual small country is a very minor player in terms of world trade in grain. The rest of the world may be supplying hundreds, if not thousands, of times more grain onto the world market than a small country.

When we are considering the trade relations of a small country with the rest of the world, we find that a different situation arises. We can illustrate this by imagining that countries A, B and C become very much larger than country D. Let us expand the size of each of these three countries by 100. This makes country D a small country. Table 7.8 and Fig. 7.8 show the new situation.

The reader will observe that in the previous figure, the curve describing the rest of the world's demand for imports was sloping downwards to the right. In Fig. 7.8, the same curve is not sloping downwards; it is flat and horizontal. This is because, as we have drawn it, the *x*-axis in Fig. 7.8 extends only to a quantity of 80 bushels. If we were to redraw the figure so that the *x*-axis extended to a quantity of 7500 bushels, which is the quantity of imports that the rest of the world demands when the price is $1 per bushel (from Table 7.8), the downward slope of the curve would be evident. However, stretching the *x*-axis to a quantity of 7500 bushels would mean the vertical axis obscuring the two curves that describe D's supply of

Table 7.8. Country D as a small country in the context of the rest of the world (bushels).

	Country D				The rest of the world			
Price ($ per bushel)	Supply	Demand	Supply of exports	Demand for imports	Supply	Demand	Supply of exports	Demand for imports
(1)	(2)	(3)	(4)	(5)	(6)	(7)	(8)	(9)
10	80	40	40		24 000	18 000	6 000	
9	80	45	35		24 000	19 500	4 500	
8	80	50	30		24 000	21 000	3 000	
7	80	55	25		24 000	22 500	1 500	
6	80	60	20		24 000	24 000	0	0
5	80	65	15		24 000	25 500		1 500
4	80	70	10		24 000	27 000		3 000
3	80	75	5		24 000	28 500		4 500
2	80	80	0	0	24 000	30 000		6 000
1	80	85		5	24 000	31 500		7 500

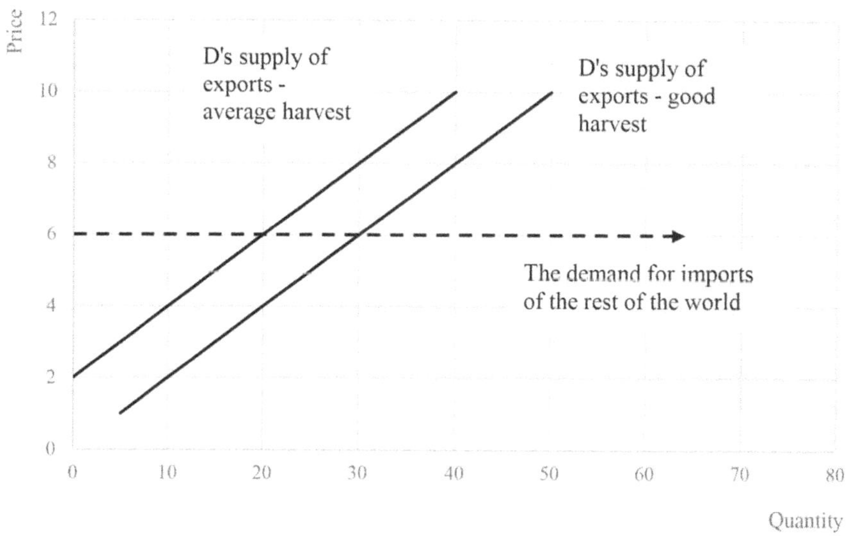

Fig. 7.8. When country D is a small country it has no discernible effect on the world price.

exports because country D's exports are so small relative to the size of the world market.

The important point that this figure illustrates is that, in becoming a small country, D no longer affects the world price to any discernible degree. The figure shows that D has expanded its supply of exports (it has shifted to the right). This is due to its better-than-average harvest. However, over the range of quantities that are relevant to country D, this has not affected the world price in any discernible manner. The world price remains at $6 per bushel. This is the case if the harvest is good or if it is bad or if it is average. Whatever the magnitude of the harvest – and the consequent demand for import and supply of exports – the world price remains, in effect, at $6 per bushel.

In contrast, when D was a large country – as in Fig. 7.6 – a good harvest caused the world price to fall (and a bad harvest caused the world price to rise). Now that D is a small country, the same good harvest has no discernible effect on the world price. The world price is, to all intents and purposes, not affected by the size of the harvest of a small country. In economic terms, when we are considering a small country that exports at the world price, the demand for imports of the rest of the world from that country is perfectly elastic with respect to price. Similarly, when we are considering a small country that imports at the world price, the supply of exports of the rest of the world to that country is perfectly elastic with respect to price.

This observation is important: in the following chapter we explore the ways in which governments of an open economy can keep the price of grain on their national markets within their price band when the world price is either above or below the price band. We shall see that governments can choose one of two approaches: either a buffer stock or trade measures (being taxes or subsidies on imports and exports). If the country is a small country, its actions will not affect the world price to any discernible degree (although if a number of small countries simultaneously takes the same action, they collectively have the effect of a large country). However, if the country is a large country, its actions do affect the world price. This distinction between small countries and large countries is comparable to a firm's situation in economic theory. A small firm is an atomistic competitor; a large firm is an imperfect competitor. The small firm is a price taker and can have no effect on the prices it faces by altering its purchases and sales; the large firm can.

7.9 Conclusion

We have developed a model of the international grain trade. It reveals the relationship between an individual country and the rest of the world. The model enables us to explore what happens when a country opens itself up to trade. For instance, does trading result in less or more price volatility? Does the price of grain within a country rise or fall when it engages in trade? The model allows us to examine how an open economy can keep the price of grain within its price band.

Notes

[1] It is notable that in the process of excluding trade within the 'rest of the world' the model generates linear supply and demand curves for its trade with Country D.

[2] Column 5 in Table 7.7 is blank because, over the range of prices considered, in a year of good harvest there is no demand for imports. Over the whole range of prices, country D is an exporter of grain.

Reference

Ricardo, D. (2004) *The Principles of Political Economy and Taxation.* Dover Publications, Mineola, New York.

8 Stabilizing Grain Prices in an Open Economy

8.1 Introduction

The purpose of this chapter is to put into use the model of the world grain market that we have developed in the previous chapter. We shall use it to explore how an open economy can stabilize the price of grain on its national market and what economic consequences may flow from this. After an explanation of why the chapter is selective in the cases that are discussed and illustrated (8.2), the chapter points out that international trade is a double-edged sword; it can both make things better for a country but it can also make things worse. Section 8.3 explains how trade can make things better. It shows how a closed economy can avoid price volatility by opening itself up to international trade. In this case, trade is beneficial. Section 8.4 explains the opposite: how trade can make things worse for a country. It shows how a closed economy with stable prices can find itself having to cope with unstable prices if it engages in international trade. In this case, trade is detrimental. Whether trade is beneficial or detrimental for a given country depends on whether the world price of grain lies within or outside the country's price band.

Is there a way for a country to engage in international trade, with all the fluctuations of the world price that this involves, and at the same time keep the price of grain on its home market inside its price band? Can a country enjoy the benefits of trade and simultaneously have a stable grain price? This is, indeed, possible. There are two approaches to the stabilization of the grain price on the national market. Governments can set up a buffer stock. We discuss this in Section 8.5. Alternatively (or in addition), governments can deploy 'trade measures'. These are subsidies, taxes and bans placed on imports and exports. They are explained in Section 8.6.

We then look at the effect that buffer stocks and trade measures have on the world as whole. A buffer stock is 'globally friendly' because it helps to stabilize the world price (8.7). On the other hand, trade measures cause 'collateral damage' to the rest of the world because they destabilize the world price (8.8). This substantiates the point made in Chapter 3 (Section 3.7), that sudden and unexpected increase in grain prices are often associated with governmental trade measures. The chapter draws its conclusions in Section 8.9.

8.2 Only a Selection of Situations Are Discussed and Illustrated

A country that trades with others is either an importer or an exporter. It can have a price problem either because the world price of grain rises above Pmax or because it falls below Pmin. Four different situations can therefore arise:

1. importer + world price rises above Pmax
2. importer + world price falls below Pmin
3. exporter + world price rises above Pmax
4. exporter + world price falls below Pmin.

If the government were to choose to use trade measures to stabilize the price on its national market, the specific measure would depend on the situation, as shown below:

1. importer + world price rise, requires an import subsidy
2. importer + world price fall, requires an import tax
3. exporter + world price rise, requires an export tax
4. exporter + world price fall, requires an export subsidy.

This chapter does not discuss and illustrate all four situations. Neither does it discuss the mechanics and the economic consequences of all four trade measures. That would require a longer discussion and a proliferation of very detailed figures, each one illustrating a specific situation or a specific measure. In the interests of readability, we restrict ourselves to a selection of situations and measures. They are intended to present the argument and serve as examples of the principles involved.

8.3 Trade Can Bring Price Stability

In this and the following four sections, we assume that the country under consideration is a 'small' country – small in terms of the volume of grain that it trades relative to the volume traded by the rest of the world. As pointed out in the previous chapter, a small country is a price taker; it faces a supply of exports or demand for imports from the rest of the world that is perfectly elastic. In other words, the rest of the world's supply of exports to the country and its demand for imports from the country is a flat, horizontal line. Whether the country increases or decreases its exports or increases or decreases its imports makes no difference to the world price. This is because the change in the volume of imports and exports is negligible compared to the volume of trade that is taking place in the world as

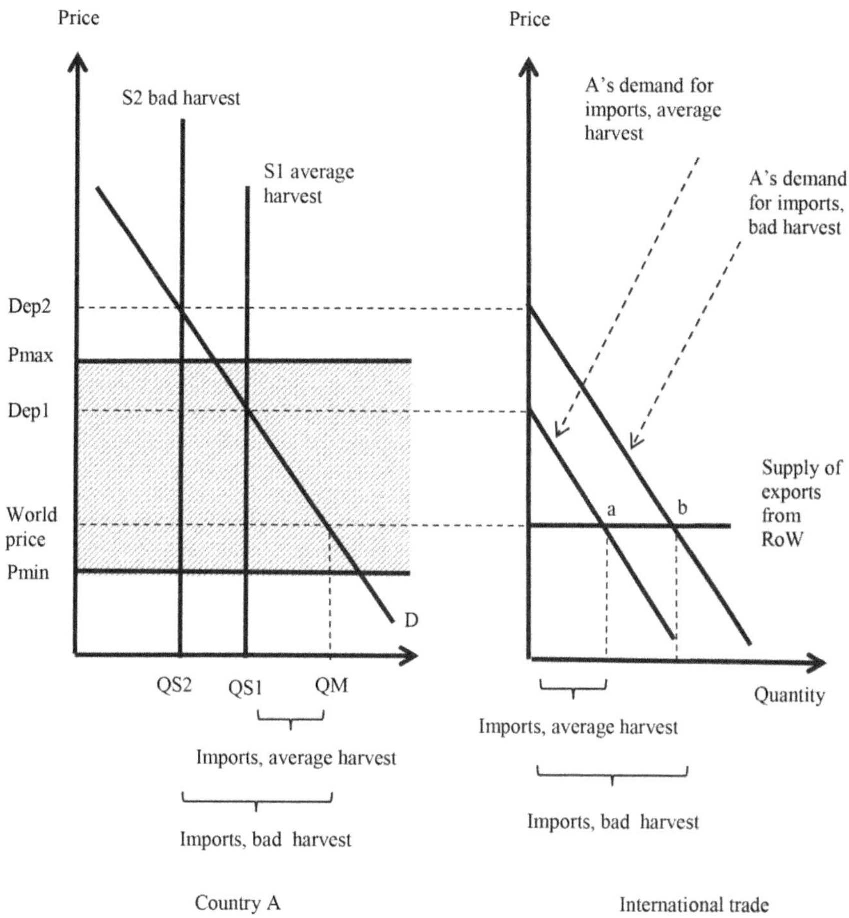

Fig. 8.1. Trade can bring stable prices.

a whole. In Section 8.8, where we consider the effects that trade measures have on the rest of the world – more specifically, on the world market price – we shall relax this assumption. Instead of considering a small country, we shall consider a 'large' country. When a large country changes its volume of exports and imports, it <u>does</u> have an effect on the world price. In this case the curves that describe the rest of the world's supply of exports and demand for imports are not flat, horizontal lines. They slope upwards or downwards, as the case may be.

An open economy buys and sells grain at the world price. So long as the world price remains within the price band, the price on the national market will also remain within the price band, irrespective of whether the harvest within the country is good, bad or average. Figure 8.1 illustrates this point.

Let us explain Fig. 8.1. Its purpose is to show that in an open economy, the price of grain within the country can be more stable than in a closed economy. In other words, if a country opens itself to trade, the price of grain on its national market can be more stable than would be the case if the country were a closed economy. This is because the world price can be more stable than the country's domestic equilibrium price.

The left-hand panel describes the market in country A; the right-hand panel describes international trade. Let us begin with explaining the left-hand panel. We start with a year of average harvest in country A. Supply is S1, giving a quantity that farmers sell on the market of QS1. If there was no trade, the market price would be the country's domestic equilibrium price, dep1. This price is within the country's price band – it is affordable to consumers. In a year of average harvest there is no price problem. The country would consume quantity QS1, which its consumers would buy at the price of dep1.

The price of grain on the world market is, however, cheaper than dep1. This is shown in the left-hand panel. It makes sense for the country to trade and to gain access to the world market because grain from the world market is cheaper. Because it is cheaper, consumers demand a greater quantity than QS1. They demand QM. There is a gap of QM less QS1. This gap is filled by imports, as shown in the left-hand panel.

Turning now to the right-hand panel, we note that supply of exports from the rest of the world (RoW) to country A is perfectly elastic, since A is a small country. The demand that country A expresses for imports in a year of average harvest is shown. The point at which the supply of exports intersects with the demand for imports is shown at point a. This denotes the quantity of grain that country A imports. This quantity is shown as 'imports, average harvest'. It is the quantity of grain that fills the gap of QM less QS1 shown in the left-hand panel. This is the situation in a year of average harvest. It does not pose a problem for country A because the world price is within its price band.

Now, let us consider what happens if the harvest in country A is bad and if country A does not engage in trade with the rest of the world. Supply is S2. The bad harvest has caused supply to shift to the left. Farmers sell a quantity of QS2. Demand for grain within A remains the same (D). Without trade, in other words without any imports, the price on A's market escalates to the domestic equilibrium price, dep2. This is higher than dep1 because the harvest is bad and there is less supply. Moreover, it is above Pmax and too expensive for consumers. There is now a price problem.

Country A can solve this problem by engaging in trade with the rest of the world. As shown in the right-hand panel, the bad harvest has generated a new demand-for-imports curve. The new curve intersects with the supply of exports from the rest of the world at point b. Country A now imports a larger quantity of grain, as shown in the right-hand panel. This quantity of imports fills the gap QM less QS2.

The upshot is that if country A engages in trade a bad harvest does not adversely effect consumers. Although supply of grain contracts from S1 to S2, meaning that the country has produced less grain, it has averted a shortage by importing grain from the rest of the world. In this particular case, trade not only provides the physical grain to fill the domestic gap but it also provides it at a price that lies within the price band. By opening itself to trade, country A avoids the vagaries of its own harvests and stabilizes the price for its consumers. Trade provides an automatic solution to a bad harvest.

An equivalent argument can be made regarding good harvests. A good harvest shifts the supply curve to the right. In this instance, trade disposes of the surplus (in the form of exports) that would otherwise remain on the market of country A and cause the price to fall below Pmin. By acting as an automatic shock absorber of variable harvests – exporting grain and importing grain as necessary – trade can stabilize the price for both the farmers and the consumers of country A.

8.4 Trade Can Bring Price Instability

8.4.1 An importing country and a rise in the world price

The fortunate situation described above – the world price for grain being within the price band of country A – does not always pertain. Trade brings stable prices only when the world price lies within the country's price band. This cannot be guaranteed. As we have shown in Chapter 3 on the pattern of grain prices, the world price is far from stable; it can rise above a country's price band; it can equally well fall below it. In this section we will show how the instability of the world price of grain can cause difficulties both for countries which import grain and for those which export it.

Figure 8.2 shows how a bad harvest somewhere in the rest of the world (RoW) can cause a problem of high prices in a country that relies on grain imports.

The purpose of Fig. 8.2 is to show how events in the rest of the world – such as a bad harvest – can drive up the world price of grain. This can cause a shortage of food in a country which relies on imports to feed its population. We saw in Chapter 2 that three quarters of the world's countries rely on imports of grain. A rise in the world price of grain can affect a country's population – in the form of higher food prices – even though the harvest of that country is average. A bad harvest in a given country is not required for the price of food in that country to escalate and become unaffordable. It is enough for there to be a bad harvest somewhere else in the world. Thus a food shortage – in the form of unaffordable prices – can occur in a country even though that country's own harvest is average. The point is to explain how countries which are open economies expose themselves to the risk of price instability. They risk prices rising too high and falling too low

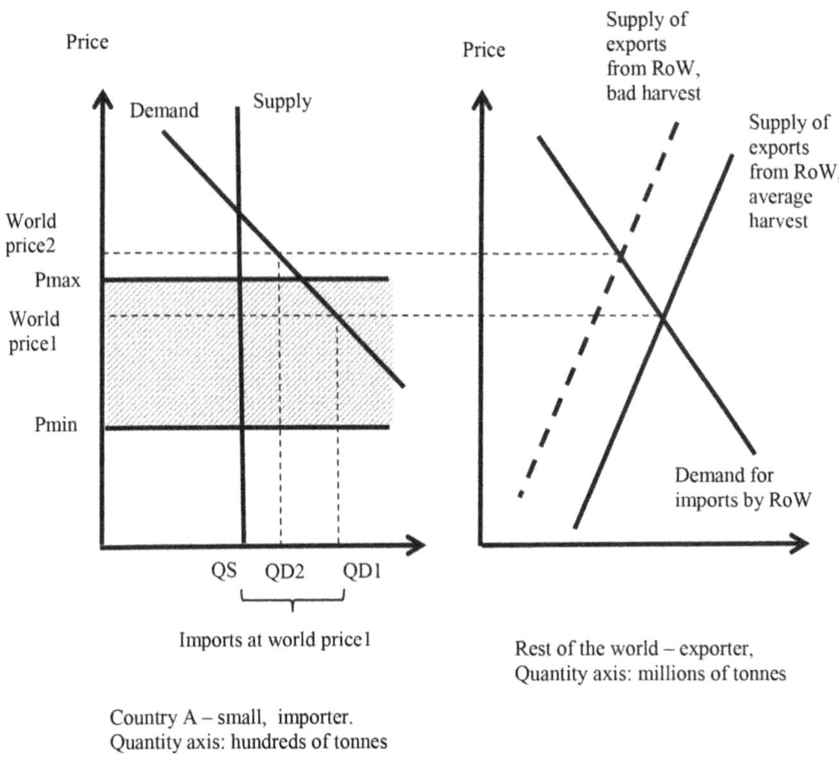

Fig. 8.2. Effect on an importing country of a bad harvest in the rest of the world (RoW).

due to events – droughts, floods, export bans and so forth – taking place elsewhere in the world and completely beyond their control.

We begin with the right-hand panel. This describes the market in the rest of the world. It shows the supply of exports from all those countries that are exporters at the world price. It also shows the demand for imports from all those countries, except country A, which are importers at the world price. These two forces balance each other at world price1.

Turning to the left-hand panel, this describes the market of country A. Supply is S, giving a quantity that is sold by farmers of QS. Demand is D. The country is open to trade and buys grain from the world market at the world price of world price1. The gap between home production and consumer demand is met by imports. The quantity of imports is QD1 less QS.

The situation described is satisfactory from the point of view of country A. The grain that it does not buy from its own farmers it imports. The price of all grain is world price1, which is within the country's price band. The population can afford to feed itself and there is no price problem.

Now let us suppose that there is a bad harvest in a grain-exporting country somewhere else in the world. This is shown in the right-hand panel by the dashed curve. The bad harvest means that the supply of exports from the rest of the world has contracted because there is less grain to export over the range of prices considered. This is represented by the shift of the curve to the left. Demand for imports by the rest of the world remains the same – that curve does not shift. But the contraction of supply of exports causes the world price to rise, to world price2. This has an impact on all other countries that engage in trade.

The impact on country A of the rise in the world price is shown in the left-hand panel. The price of all grain – including home-grown grain – rises from world price1 to world price2. The quantity demanded decreases from QD1 to QD2. The price is now above Pmax. The country has a price problem, even though its own farmers did not decrease the quantity of grain that they produced (QS remains unchanged). The problem has arisen not because the country's harvest was bad (it was an average harvest) but because of events elsewhere in the world. There was a bad harvest in another country – an event over which the government of country A had no control.

This is a case of a rise in the world price above Pmax causing the price of food in an importing country to become unaffordable. The opposite situation, which is not shown, can equally occur: the world price may fall below Pmin, due to, say, a very good harvest elsewhere in the world. In that case, it is not the consumers who are hit, it is the country's farmers because the price at which they sell their grain is too low to return an acceptable profit. In other words, an importing country can suffer a price problem both when the world price rises and when it falls.

8.4.2 An exporting country and a fall in the world price

Let us now turn from an importing country to an exporting country. Ostensibly, an exporting country is in the fortunate position of having a surplus of grain. It produces more grain than its consumers demand. It exports the surplus and, at least in principle, earns money in so doing. It may be thought that a country that can export grain is in a much better position than a country that needs to import grain. This is not necessarily so. Fluctuations in the world price of grain inflict similar problems on exporting countries as they inflict on importing countries. Figure 8.3 shows the situation of a good harvest somewhere in the rest of the world (RoW).

Let us start with the left-hand panel. This describes the market in country A. The world price is world price1. The initial situation in country A, an open economy, is that its farmers sell quantity QS. At world market1, the quantity demanded by the consumers of country A is QD1. The difference between the quantity supplied by the farmers (QS) and the quantity demanded by the consumers (QD1) is exported at world price1. Since this price is within the country's price band, there is no price problem. Consumers can afford to buy food and the farmers receive an acceptable return for their work. It is a satisfactory situation for country A.

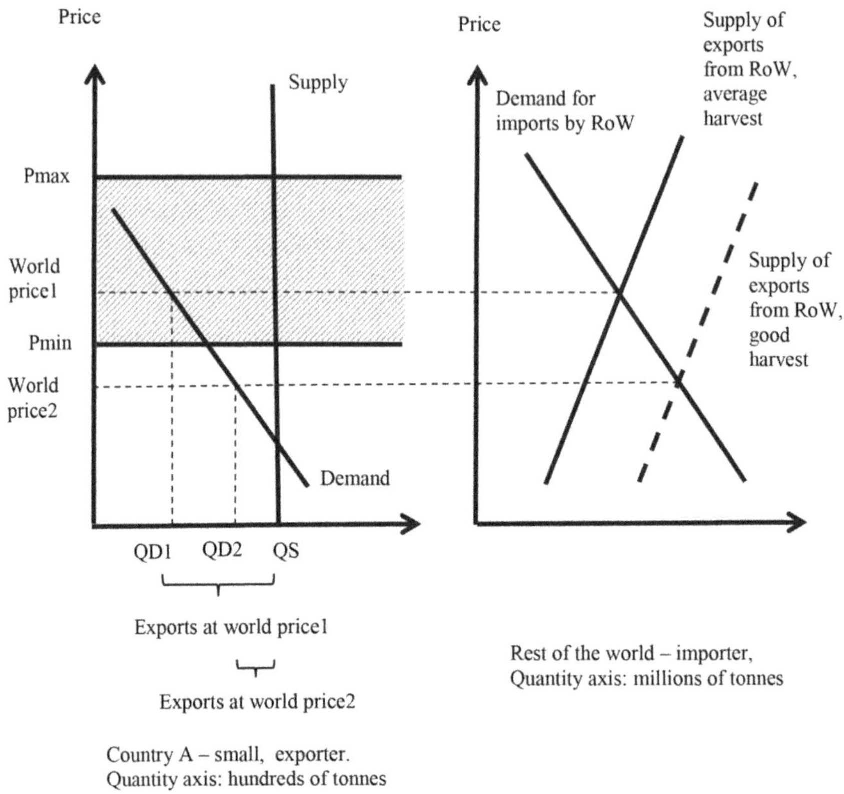

Fig. 8.3. Effect on an exporting country of a good harvest in the rest of the world (RoW).

Let us turn to the right-hand panel. Suppose that elsewhere in the world there is a good harvest (i.e. bigger than average) in an exporting country. This causes the supply of exports from the rest of the world to expand. The new supply-of-exports curve is the dashed line. The world price falls to world price2. At this lower price, consumer demand in country A expands to QD2, in the left-hand panel. The quantity of exports falls from the previous quantity of QS less QD1 to a smaller quantity of QS less QD2. A price problem has arisen: farmers sell their grain at world price2, which is less than Pmin. This problem has arisen not because the farmers have oversupplied their own market but because of events elsewhere in the world – and beyond the control of the government of country A. Due to a good harvest in another country, the farmers of country A cannot earn an acceptable living.

This is the case of a fall in the world price causing a price problem for the farmers of an exporting country. The opposite situation, which is not shown here, can equally well occur: a rise in the world price can cause problems for the consumers of an exporting country. A rise in the world market price can be the consequence of a bad harvest somewhere in the world. If the world price rises above Pmax, this bring a price problem for consumers. A rise in the world price affects both consumers in countries that import grain and consumers in countries that export grain. We have thus seen how exporting countries are not necessarily in a more fortunate position than importing countries. All countries that engage in international trade are exposed to volatile world prices.

8.5 The Use of a Buffer Stock to Stabilize Prices

Our three figures (Figs 8.1–8.3) illustrate a vital point: that free trade is sometimes, but not always, a panacea for unstable grain prices. Figure 8.1 shows that when the world price is within a country's price band, free trade is a solution. The other two figures show that when the world price departs from the price band, free trade becomes a problem. Clearly, free trade cannot always bring stable prices. By its very nature, free trade is not a permanent solution to price volatility. The question is, how can a country engage in trade and, at the same time, enjoy a stable grain price?

As stated in the introduction to this chapter, the government of an open economy has two options to keep the price of grain on its national market within its national price band. It can either set up a buffer stock or it can deploy trade measures. If the government chooses the buffer stock approach, it places itself in a position to offer to sell grain to its consumers whenever the world price of grain rises to Pmax. This prevent the price of grain on the national market from rising above Pmax. It avoids a price problem for consumers. The price of grain on the world market will, however, be higher than Pmax. Traders will find that they can make a profit by buying grain in the country and selling it on the world market. This would drive the price on the country's own market above Pmax. It would defeat the government's intention. To avoid this, the government needs to place a temporary tax on exports.[1] An export tax prevents grain being sucked out of the country due to the higher world market price. In this way, a buffer stock can prevent a price problem for consumers.

What about the opposite situation – a fall in the world market price? Can a buffer stock avert a price problem for farmers? By setting up a buffer stock, the government is in a position to offer to buy grain from its farmers whenever the world price of grain falls to Pmin. The government deposits the grain that it has purchased at Pmin from its farmers in the buffer stock, thereby replenishing it. By buying grain at Pmin, the government prevents the price on its home market from falling below Pmin. This avoids a price problem for the country's farmers. While the price in the country will be at Pmin, the price on the world market will be lower. It will be profitable for traders to buy grain on the world market and sell it in the country. The country does not want this to happen because it will drive down the price on the country's market. If the government is to stop the price falling below Pmin, it will find itself having to buy up an unlimited flow of imported grain. To avoid grain being sucked into the country from the world market, the government needs to place a temporary tax on imports.[2] This will mean that it is not profitable for traders to buy grain on the world market and import it into the country.

In essence, the government of an open economy can deploy a buffer stock in exactly the same way as the government of a closed economy, as we have explained in Chapter 6: offering to sell grain to consumers when the price rises to Pmax and offering to buy grain from farmers when the price falls to Pmin. A closed economy does not trade with the rest of the world on a permanent basis while an open economy does trade with the rest of the world, except during those periods when the world price of grain rises above Pmax or falls below Pmin. During these periods, an open economy needs to disengage, on a temporary basis, from trading with the rest of the world. At all other times, i.e. when the world price of grain lies within a country's price band, it can import grain from, or export it to, the rest of the world. The country stops trading with the rest of the world only during those periods when the government is selling grain from its buffer stock to its consumers, or buying grain into its buffer stock from its farmers. If it did not curtail trade during these periods, grain would be sucked out of the country and onto the world market at times of high prices, and sucked into the country from the world market at times of low prices. This would defeat the purpose of the buffer stock.

8.6 The Use of Trade Measures to Stabilize Prices

8.6.1 Averting a rise in the market price with an import subsidy

The alternative to a buffer stock – or as a complement to a buffer stock – is 'trade measures'. As stated in the chapter's introduction, there are four different trade measures: import subsidies, import taxes, export taxes and export subsidies. Trade measures act as incentives and disincentives to traders, influencing the quantity of grain that traders import or export. When the level of import tax is so high as to make it unprofitable for traders to import grain into a country, the tax is a *de facto* ban on the importation of grain. The country then imports no grain. Ditto for an export tax. To avoid an over-lengthy discussion and too many technical figures, we restrict this section to a consideration of import subsidies and export subsidies. The mechanics of import taxes and export taxes are identical but the effects are the opposite.

Trade measures can keep the market price of grain within a country's price band when the world price either rises above Pmax or falls below Pmin. We consider first a rise in the world price and the use of an import

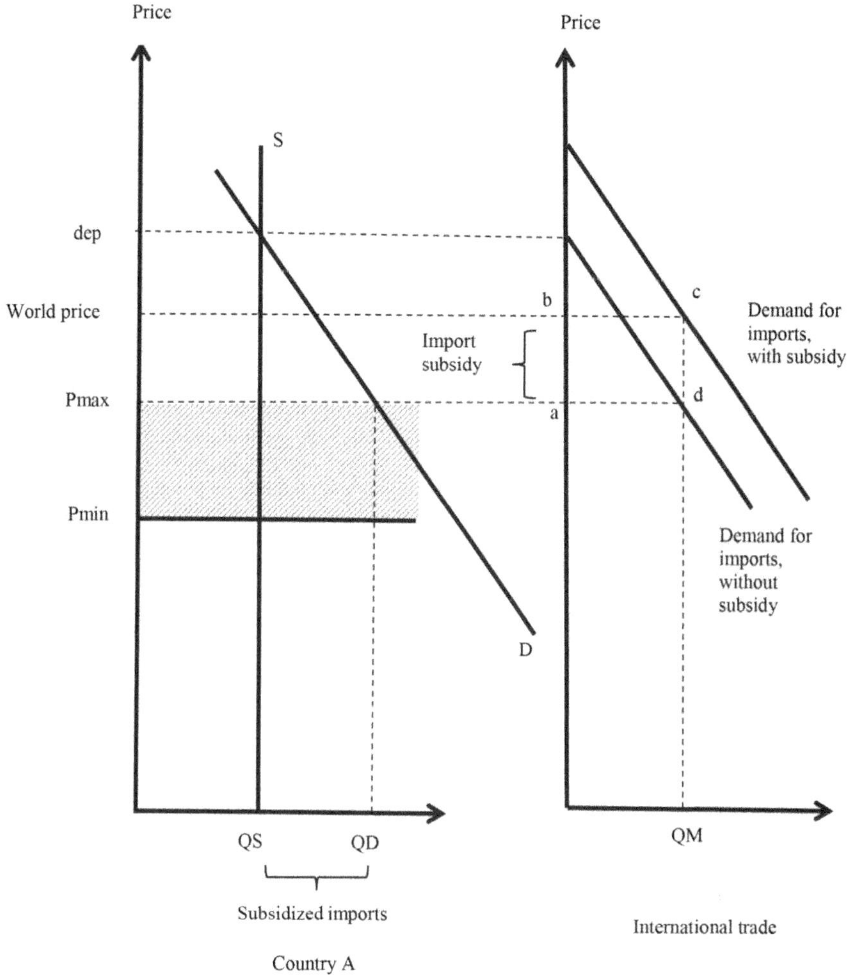

Fig. 8.4. The world price rises above Pmax, use of import subsidy.

subsidy designed to reduce the price of imported grain. This situation is described in Fig. 8.4.

In Fig. 8.4, the world price is above Pmax.[3] The task of the government is to prevent the market price within the country rising above Pmax. It can do this by introducing a subsidy on grain imports. This shifts the import demand curve to the right as shown in the right-hand panel. Traders still buy grain on the world market at the world price. When they bring it into country A they receive the subsidy from the government. In their minds, the actual price that the grain costs is the world market price less the government subsidy. This causes traders to buy a greater quantity than they would buy in the absence of the subsidy (for an explanation of how a subsidy changes the purchasing behaviour of consumers, see Section 6.5 on mitigation measures).

With the subsidy in place, traders import a quantity of grain equal to QM, shown in the right-hand panel. When this quantity of grain is added to the quantity supplied (QS) by the farmers of country A, the total quantity supplied on the market is QD. At this level of supply and, given the quantity demanded (QD), the market price within the country is Pmax. A price problem has been averted because the price on the market of country A is now within its price band. We should note, however, that Pmax is lower than the world price. There is, therefore, an incentive for traders to buy grain in country A at Pmax and export it to the world market. If they did this, they would make a profit. It would mean that grain that had been imported would now be exported. This would undermine the whole purpose of the import subsidy. To prevent grain being re-exported, the government needs to introduce an export tax. This removes any incentive to traders to re-export grain.

There is, of course, a cost to the government. It has to pay the subsidies to the traders who import the grain from abroad. The unit subsidy (per tonne) is the difference between the world price and Pmax. The government pays the subsidy on all imported grain (QM). The cost to the government is the world price less Pmax multiplied by QM. This is the area abcd shown in the right-hand panel.

8.6.2 Averting a fall in the market price with an export subsidy

Having considered a rise in the world price, let us now turn to a fall in the world price and see how a government can avert a price problem for farmers by means of a trade measure. In this case, the world price has fallen below Pmin.[4] The government can introduce an export subsidy. This is described in Fig. 8.5.

From the left-hand panel, we see that the domestic equilibrium price (dep) is below the world price. This means that it is profitable for traders to arbitrage grain by buying grain in country A and exporting it to the rest of the world. Traders will continue to do this until the price in country A is equal to the world price. Country A is therefore an exporter of grain, exporting the quantity QX1 (in the right-hand panel), which is the same as QS less QD1 in the left-hand panel. However, even though the country is exporting grain, its farmers have a price problem: the price at which they sell grain (the world price), whether they sell it for export or to domestic consumers, is below Pmin.

To solve the farmers' problem, the government can introduce a subsidy on exports of grain. This shifts the country's export supply curve to the right, shown in the right-hand panel. Traders buy grain in country A at the world price, as they did before the government provided the subsidy. As traders move the grain over the border to sell it on the world market,

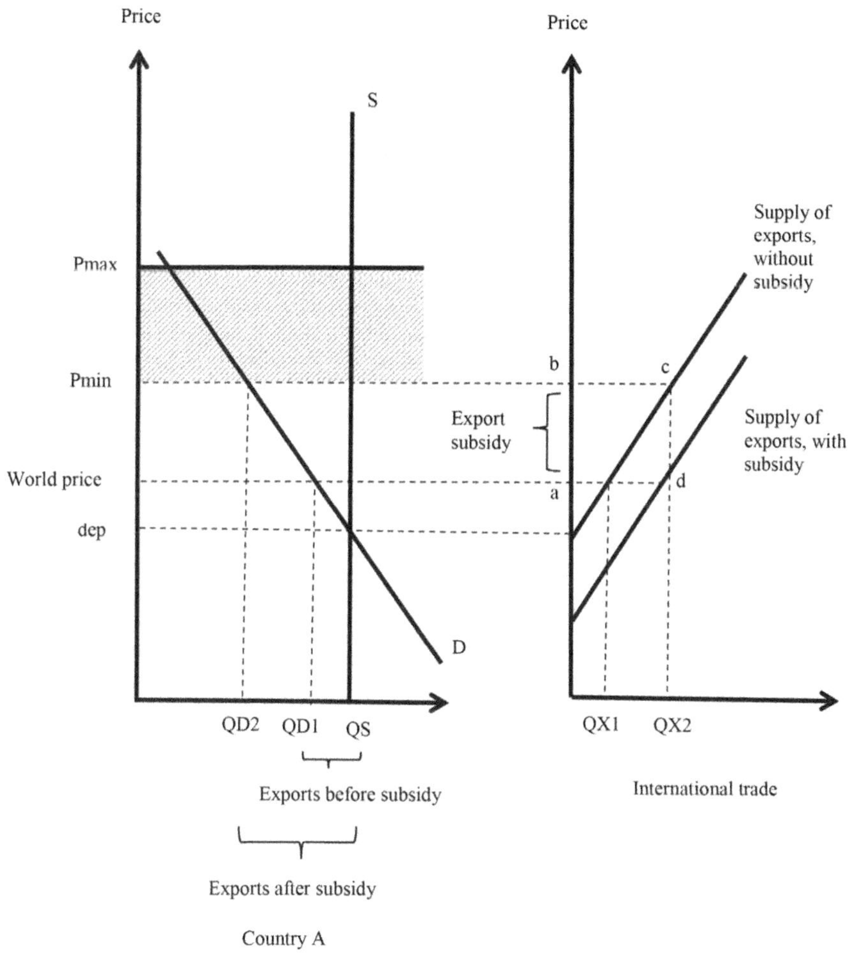

Fig. 8.5. The world price falls below Pmin, use of export subsidy.

the government pays them the subsidy. In the minds of the traders, the actual price that they pay for grain when they purchase it in country A is the world price less the government subsidy. This induces traders to increase the quantity that they buy (see Section 6.5). Traders now export a quantity of QX2. This is the same quantity as QS less QD2 in the left-hand panel. When this quantity of grain is removed from the country, the quantity of grain remaining in the country is QD2. In effect, the export subsidy has brought about a larger quantity of exports and has moved the supply onto the market of country A to QD2. At this level of supply, and given the quantity demanded of QD, the price is Pmin.

Since Pmin is higher than the world price, the government needs to place an import tax on grain. If it does not, grain is sucked into the country from the world market. This would undermine the purpose of the export subsidy which is to induce traders to export more grain from country A than they otherwise would do. The import tax removes any incentive to traders to import grain into country A.

As with the import subsidy above, there is a cost to an export subsidy. The unit cost (per tonne) is Pmin less the world price. This subsidy is paid on all grain exported. This is the quantity QX2, shown in the right-hand panel. The cost that falls to the government's budget is therefore Pmin less the world price multiplied by QX2. This is shown in the right-hand panel as the area abcd.

8.7 The Effect of Buffer Stocks on the World Price: 'Globally Friendly'

Having exposed the mechanics of buffer stocks and trade measures we now turn to their broader economic consequences. If the government of a country sets up a buffer stock and uses it to stabilize the price of grain on its home market, does this affect other countries? If it does affect other countries, is this for good or for ill? Likewise, when a government deploys a trade measure, does this have a broader impact?

A buffer stock can keep the price of grain on the home market within the country's price band. Over and above this, however, a buffer stock has an impact on the world market price. The mechanism is as follows. If the world price of grain is rising, the task of the government is to prevent the price on its home market from rising above Pmax. It does this by selling grain from its buffer stock to its consumers at the price Pmax. Let us assume that the country is an importer. By selling grain from its buffer stock to its own consumers, the country will import a smaller quantity than it would in the absence of a buffer stock. Grain from the buffer stock supplants imported grain. This means that there is more grain on the world market than would otherwise be the case. More grain on the world market means that the rise in the world price is mitigated. By selling grain from its buffer stock to its own consumers, the government relieves some of the upward pressure on the world price. At a time of rising grain prices, this is beneficial for the rest of the world. From this point of view, a buffer stock is 'globally friendly'.

Let us take the opposite situation: the world price is falling. Suppose the country is an exporter of grain. To prevent the price of grain falling below Pmin on its home market, the government offers to buy grain from its farmers to deposit in the buffer stock. In buying grain from its farmers, the government brings about an expansion of demand. This expansion of demand means that the country's supply of exports is less than would otherwise be the case. This, in turn, means that there is less grain offered on the world market. The downward pressure on the world price is relieved.

In other words, by buying up grain into a buffer stock, the government exports less, which means that there is less downward pressure on the world price. At a time of falling grain prices, this is beneficial for the rest of the world. Again, from the point of view of the rest of the world, a buffer stock is 'globally friendly'.

8.8 The Effect of Trade Measures on the World Price: Collateral Damage

8.8.1 An import subsidy causes the world price to rise

We have shown that a buffer stock is beneficial to the rest of the world. This is in stark contrast to trade measures. They have a malign effect on the rest of the world. They cause 'collateral damage', by destabilizing the world price. In so doing, they are one of the main causes of the problem of price volatility itself. In curing the problem of volatile prices with trade measures, a country makes the disease worse for everybody else. This picks up a point made earlier. In the discussion in Chapter 3 on the pattern of grain prices, we proposed that there are three factors behind sudden surges in the price of grain: bad weather during the growing season; a low level of world grain stocks; and government trade measures (see Table 3.1).

To illustrate how trade measures destabilize the world price, we pass from considering the case of a small country to a large country. This is not to suggest that the actions of small countries do not have any effect on the world price; they do, but the effect of a small country on the world price is imperceptible compared to the effect of a large country. This is because, as we have explained in Section 7.8, we can consider that the supply of exports (and demand for imports) by the rest of the world to/from a small country to be a flat horizontal line, while the supply of exports (and demand for imports) to and from a large country is a line that slopes upwards. When a large country introduces a trade measure, we can expect that this will alter the world price. When a small country introduces a trade measure, the effect on the world price will be imperceptible. However, we should remember that if all (or a group of) small countries act in the same manner, they have the same effect as a large country. To see how trade measures cause collateral damage, we first examine the effect of an import subsidy on the world price, followed by the effect of an export subsidy on the world price.[5]

Figure 8.6 describes how an import subsidy causes the world price to rise.

In Fig. 8.6, the world price is at P1.[6] Country A has a price problem because P1 is higher than Pmax. The government's task is to prevent the price of grain on its national market rising above Pmax. It does this by introducing a subsidy on grain imports, as we have described in Fig. 8.4. This shifts the demand for imports to the right, shown in the right-hand

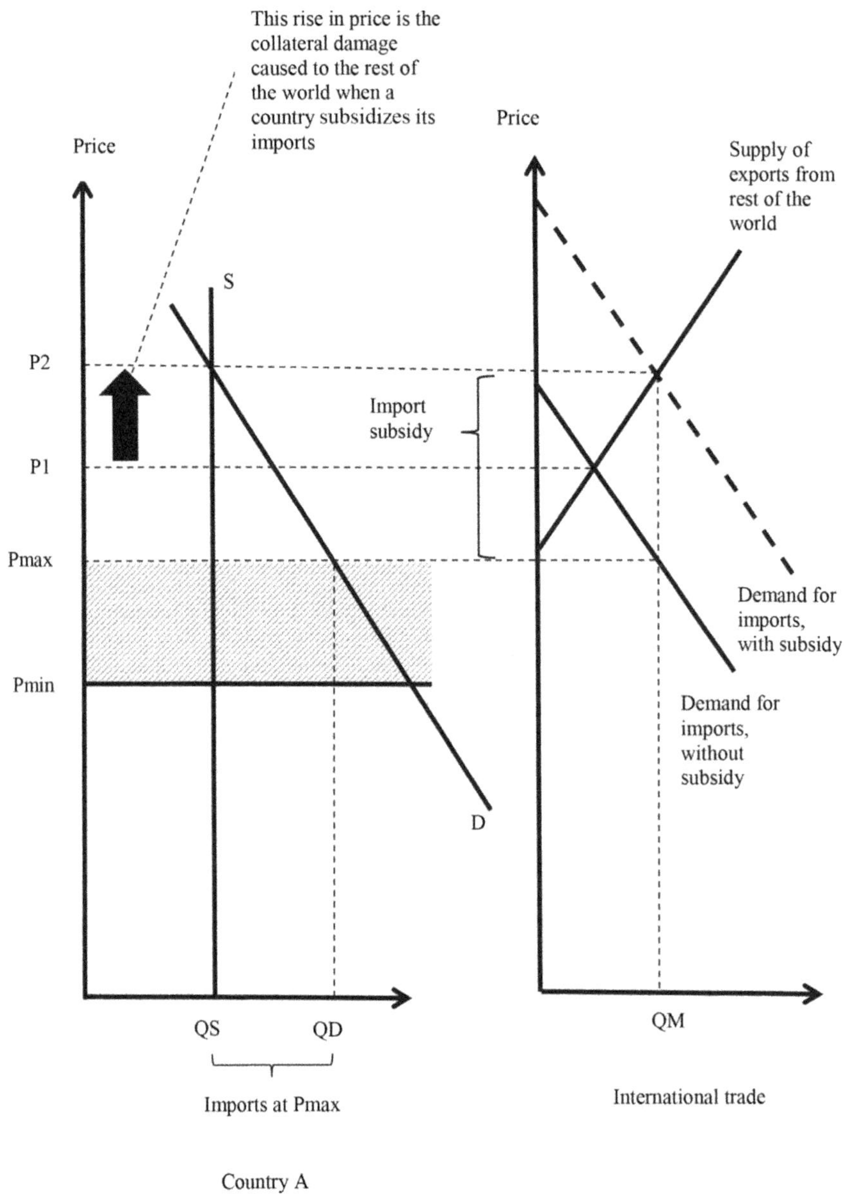

Fig. 8.6. Import subsidy causes world price to rise.

panel by the dashed line. The new world price is P2. This is the price that now prevails in the rest of the world. It is the price at which grain is traded between countries and is also the price that prevails on the domestic market of all open economies other than country A. Consumers in all other countries find themselves having to pay a higher price for their grain-based foods. The rise in the world price of grain for all other countries is the collateral damage wrought by the introduction of an import subsidy by the government of a country.

Some readers may remark that although an import subsidy may be deleterious for consumers, it may be good for farmers. The price of food for consumers will rise, but so also will the price at which farmers sell their grain. Farmers will be better off. The harm to consumers will be offset by the benefit to farmers. Is it not a zero-sum game? In principle, this is correct. However, governments subsidize imports when the world price of grain is already high. At these times, the priority is to keep food affordable to consumers. The priority is not to provide farmers with higher prices.

Clearly, it is better from a global point of view if governments use buffer stocks rather than trade measures to cope with a rise in the world market price of grain. In reality, it may be the case that a country's buffer stock is too small to address a sustained and substantial increase in the world market price. The government may find that it runs out of grain to satisfy the demand of its consumers at Pmax. In that case, the government will need to import at least some grain and subsidize its sale. But the quantity that it imports will be less if there is a buffer stock in place, than it would be if there were no buffer stock.

8.8.2 An export subsidy causes the world price to fall

Having examined the economic consequence of an import subsidy on the rest of the world, we now examine the effect of an export subsidy. Does this also have a malign effect on other countries? Figure 8.7 shows the situation. (We may note that this figure is, in essence, the same as Fig. 7.5 of the previous chapter. Both figures describe a large country exporting to the rest of the world. In both figures, the demand for imports of the rest of the world is a downward-sloping line. It is downward-sloping because, as explained in Section 7.8, in terms of quantity of grain traded, the country that is exporting is 'large' relative to the rest of the world.)

We start with the left-hand panel. The world price is P1. It is less than Pmin. There is thus a price problem for the country's farmers. At the same time, the country is exporting grain at the world price. The quantity of grain that is exported is QS less QD1, in the left-hand panel. This is the same quantity as QM1 in the right-hand panel. To resolve its price problem (the price is too low for its farmers), the government of country A provides a subsidy on its grain exports. This generates a new supply-of-exports curve – the dashed line in the right-hand panel. The supply of exports from country A to the rest of the world has expanded. The line has shifted to the right.

There is now a new point of intersection of the supply of exports from country A and the demand for imports of the rest of the world. This is a new point of balance on the world market. The world market finds its

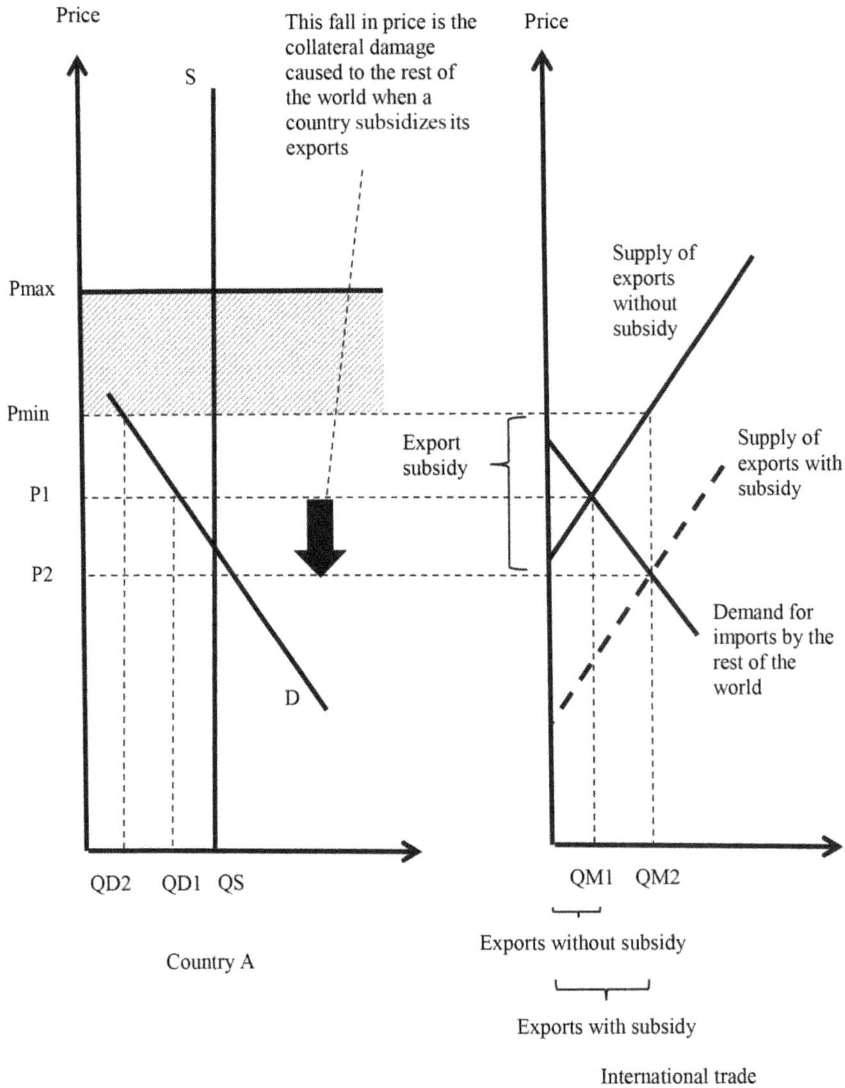

Fig. 8.7. Export subsidy causes world price to fall.

balance at a quantity of traded grain of QM2 and at a price of P2. The world price has fallen from P1 to P2. The quantity of grain exported by country A to the rest of the world has increased from QM1 to QM2.

The price problem that faced the farmers of country A is now resolved; they receive a price for their grain of P2 plus the subsidy. When these are added together, the price comes to Pmin – so these farmers make an acceptable living. But it is farmers in the rest of the world that pick up the tab. Before the government of country A introduced its export subsidy, farmers elsewhere in the world had received the price of P1 for their grain. Now they receive a lower price – P2. This is the collateral damage inflicted by country A's export subsidy on the rest of the world.

The livelihood of farmers in other countries may be placed in jeopardy. Moreover, the amount of foreign exchange that a country earns from growing and exporting grain – foreign exchange that may be essential for the development of the country – may be reduced. This is not a theoretical proposition. For several decades governments the world over have complained of the export subsidies paid by the EU and the USA on agricultural exports. They complain that subsidized exports constitute unfair competition and can put local farmers out of a job.

The reader may remark that the harm to farmers caused by a low world price is offset by the benefit to consumers of cheaper food. Is the situation not a zero-sum game? In theory, yes; but from a practical point of view, governments only subsidize grain exports when the world price of grain is low. At these times, the farmers need higher prices more than the consumers need cheaper food.

We thus see how a government – especially that of a large exporting country – by stepping in with an export subsidy to solve its own price problem, causes collateral damage to the rest of the world. As in the case of an import subsidy, it is better – from the point of view of the rest of the world – for the country to deploy a buffer stock to resolve the problem of the low grain price. This would be a 'globally friendly' approach because it would relieve some of the downward pressure on the world price. In practice, it may be that a government finds that it lacks enough spare capacity in the buffer stock to absorb the quantity of grain that needs to be purchased to prevent the price falling below Pmin. The government may not be able to buy up all the grain that it needs to purchase, for the simple reason that there is not the physical capacity to store it. In that case, the government will have to supplement its buffer stock policy by exporting some grain with a subsidy. However, the overall effect is that, with a buffer stock, the collateral damage to the rest of the world will be less than it would be if the government were to pursue a pure trade policy.

8.9 Conclusion

This chapter has examined how countries can both engage in international trade *and* enjoy stable grain prices. They can do so by either setting up a buffer stock or by deploying trade measures (subsidies or taxes on imports and exports). We believe that buffer stocks are the better approach, for two reasons. First, we have already observed in Section 3.7 how a low level of world stocks is associated with sharp rises in world grain prices. It makes sense to keep a high level of world stocks. If governments set up buffer stocks, this increases the level of stocks in the world as a whole and thus reduces the possibility of a sharp rise in the world price of grain. The second reason is that trade measures, when instigated by large countries, cause collateral damage to the rest of the world. This chapter has explained how an import subsidy causes the world price to rise and how an export subsidy causes the world price to fall. (Export taxes and import taxes, which the chapter has not discussed, have the same detrimental effect.)

Trade measures cause the world price to move in exactly the wrong direction from the point of view of all other countries. At a time of rising world prices, trade measures add fuel to the fire. The need is, rather, to douse the fire. Buffer stocks douse the fire. At a time of falling world prices, trade measures cause the world price to fall further. Buffer stocks have the opposite effect; they slow, perhaps even reverse, the downward decline in prices.

Governments need to take note of the beneficial effects of a buffer stock and the detrimental effects of trade measures on other countries. Our world is becoming ever more intertwined. Problems and challenges that were once national are becoming global. In this new context it is important that governments solve their own problem of unstable grain prices in a way that does not cause collateral damage to the rest of the world. Buffer stocks are such a way. From the point of view of the world as a whole, buffer stocks are 'globally friendly'. The next chapter examines them more closely.

Notes

1 The legality of a temporary export tax is discussed in Chapter 9 under WTO law.
2 The legality of a temporary import tax is discussed in Chapter 9 under WTO law.
3 In this example, we suppose that the world price is above Pmax, but is below the domestic equilibrium price of country A. This keeps the figure and the discussion uncomplicated. In reality, it can be the case that the world price can rise not only above Pmax (causing a price problem) but also above the domestic equilibrium price. In that case, country A is turned from an importer into an exporter. It finds not only that its consumers are having to pay very high prices for grain-based foods but also that the grain from its own farmers is leaving the country because it is exported. In response, governments can ban grain exports. This brings the price in the country down to the domestic equilibrium price and provides a measure of relief. However, the price is still above Pmax and the price problem remains. An export ban reduces the quantity of grain that would otherwise be supplied onto the world market and, thereby, puts upward pressure on the world price. It is for this reason that export bans meet with strong criticism. For instance, the World Bank (2008) is of the view that 'exporting countries should pledge not to apply export bans or prohibitive taxes for exports under any situation'.
4 In this example we assume that the world price falls below Pmin but that it does not fall below the domestic equilibrium price. In the real world, it can do so, and country A is turned from an exporter into a importer. To address this, an import tax is necessary. This will prevent the market price falling below Pmin and thus

provide a measure of relief. An export subsidy is still required to avoid the market price falling below Pmin.

[5] This is a partial analysis of the effects of trade measures on the world price. It assumes that the world price does not rise so high (or fall so low) that the country turns from being an importer into an exporter (or from an exporter into an importer). In reality, this can happen. The collateral damage of trade measures is then greater since the government prevents exports of grain onto the world market (or imports from the world market).

[6] In this figure, in order to clearly illustrate the economic consequences of an import subsidy, we have deliberately exaggerated the slope of the curve that describes the supply of exports from the rest of the world. In the real world, this curve would slope upwards but to a lesser extent than we have shown. As explained in the previous chapter, if country A was a small country it would face a supply-of-exports curve that is flat (horizontal). But when a country stops being small, what was a horizontal line begins to slope upwards. The country would have to become very large indeed, relative to the rest of the world, to slope up as steeply as shown in the figure.

References

World Bank (2008) *Preparing for the next global food price crisis*, Speech by Justin Lin, Senior Vice-President and Chief Economist at Roundtable, 17 October. World Bank, Washington.

9 Buffer Stocks – Technical and Legal Aspects

9.1 Introduction

In Chapter 6 we noted that a buffer stock allows the government to influence the domestic equilibrium price of grain. In the previous chapter we argued that buffer stocks avoid the collateral damage that is perpetrated by trade measures. Given these ostensible advantages, the purpose of this chapter is to probe further into the practical plausibility of buffer stocks.[1] We first ask whether buffer stocks are effective in helping to stabilize prices on national markets in the real world (9.2). We discuss the issue of the financial cost and whether they are financially sustainable (9.3). After describing their operations (9.4), we examine some practical aspects (9.5). We then discuss the law of the World Trade Organization regarding the measures that governments are allowed to take to stabilize the price of agricultural commodities (9.6). The question of whether the storage of grain by private traders can substitute for a public buffer stock is addressed in Section 9.7. Finally, the advantages of regional (multi-country) buffer stocks over national buffer stocks is discussed in 9.8.

9.2 Do Buffer Stocks Work in Practice?

Historical records show that the ancient Egyptians routinely stored grain as an insurance against the failure of the river Nile to flood each season. The pharaohs may or may not have known the definition of price elasticity of supply and demand, but they certainly realized that a buffer stock was politically essential and one of the foundations of an advanced civilization. More recently, buffer stocks have played an important role in stabilizing the price of rice. Dawe and Timmer (2012) recount the success of Indonesia and Thailand during the 1960s and 1970s in this regard.

As we have noted in Chapter 3, in 2007/8 there was a world food crisis. Very many countries were badly affected, but China avoided it. The government rolled out a stabilization strategy in which its national buffer stock

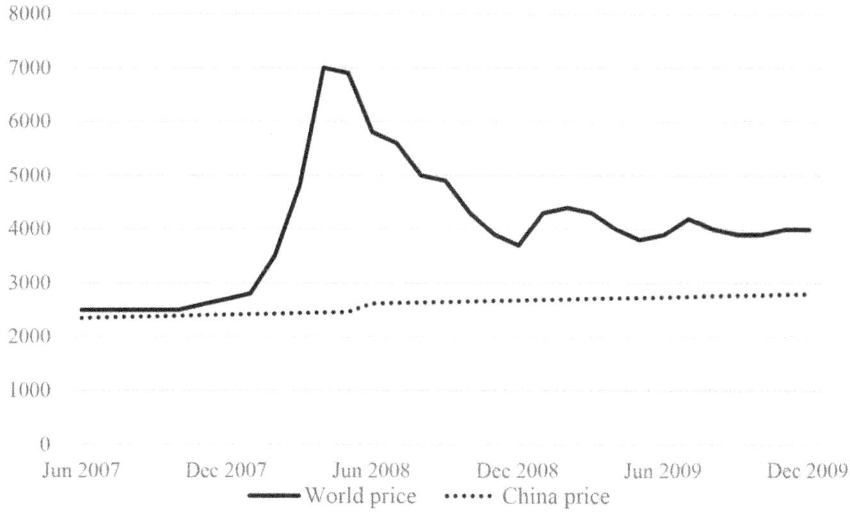

Fig. 9.1. The stability of rice prices in China (yuan per tonne). (Adapted from Huang *et al.*, 2013)

played an important role. The government was able to keep its domestic market – which had been gradually liberalized over the previous 20 years[2] well supplied with rice and wheat. The government was able to maintain stable grain prices during this crisis, while many of its neighbours were forced to buy rice at very high prices on the world market (Huang *et al.*, 2013).[3] Figure 9.1 shows that the price of grain in China remained unchanged while it exploded on the world market.[4]

In some instances, governments have set up buffer stocks with disappointing results (see, for instance, Jayne *et al.*, 2006). The reasons include a lack of clear objectives, a failure to respect operational rules regarding the price at which to buy and sell grain, and negligence by the public authorities to avoid dishonest and corrupt practices. There have been cases of serious wrongdoing. These disappointing results have led some observers to reject the entire notion of grain reserves and buffer stocks as a way of moderating the volatility of grain prices. However, this is akin to rejecting the notion of, say, air travel because of a number of air accidents. In other words, it is important to distinguish between the principles of buffer stocks and the practice of how they are managed.

9.3 The Financial Costs of a Buffer Stock

The principal cost of a buffer stock is the cost of setting it up. Storage facilities have to be built and the initial inventory of grain has to be purchased. Subsequently, the running costs can be modest and can be offset by the profits made from selling high and buying low. The fact that a buffer stock generates its own revenue means that it does not have to call on the government budget for financial appropriations. This is in stark contrast to trade measures: subsidies for imports and exports can be very expensive and have to be paid by the government. A buffer stock can be, at least in principle, financially self-sustaining.

The cost of purchasing grain to keep the price at Pmin in a year of good harvest is shown in Fig. 9.2.

In a year of good harvest, to prevent the equilibrium price from falling to P1, the government buys a quantity of grain equal to Q2 less Q1 at the price of Pmin. The cost to the government is the quantity of grain purchased multiplied by the price, shown as the shaded area Q1abQ2.

The revenue that is generated by selling grain for sale on the market is shown in Fig. 9.3.

The government sells a quantity of grain equal to Q2 less Q1. In selling this quantity, supply expands at Pmax. The new supply curve is S2 (the dashed line). The equilibrium price falls from P1 to Pmax. This keeps the price of food affordable. The amount of money that is generated by the sale of grain is the quantity (Q2 less Q1) multiplied by the price at which it is sold (Pmax). This is represented by the shaded area, Q1cdQ2. The potential profit is the difference between the purchase costs and the sales revenue. It is the difference between the two areas that are shaded in Figures 9.2 and 9.3.

9.4 Operations

9.4.1 Setting and adjusting the price band

The government needs to set a price band which straddles the long-term national equilibrium price. The maximum price (Pmax) needs to be above the long-term equilibrium price and the minimum price (Pmin) needs to be below it. If not, sooner or later the buffer stock will either run out of grain or out of money. Consider Fig. 9.4.

Let us assume that there is a buffer stock containing 70 tonnes of grain. In year 1, the domestic equilibrium price falls within the price band. No consumers wish to purchase grain from the buffer stock because the government sells this grain at Pmax, which is higher than the current market price. In year 2, the domestic equilibrium price has risen slightly, but again there are no consumers wishing to purchase grain from the buffer stock. In year 3 the domestic equilibrium price has risen again, but it remains within the price band. In year 4, the government observes that the domestic equilibrium price starts to reach Pmax. At this point, consumers will wish to buy at Pmax from the buffer stock. The government makes its grain available for sale. It continues to do so for as long as there is upward pressure on the domestic equilibrium price. Over the course of the fourth year, the government has sold 10 tonnes from its buffer stock. There remain 60 tonnes.

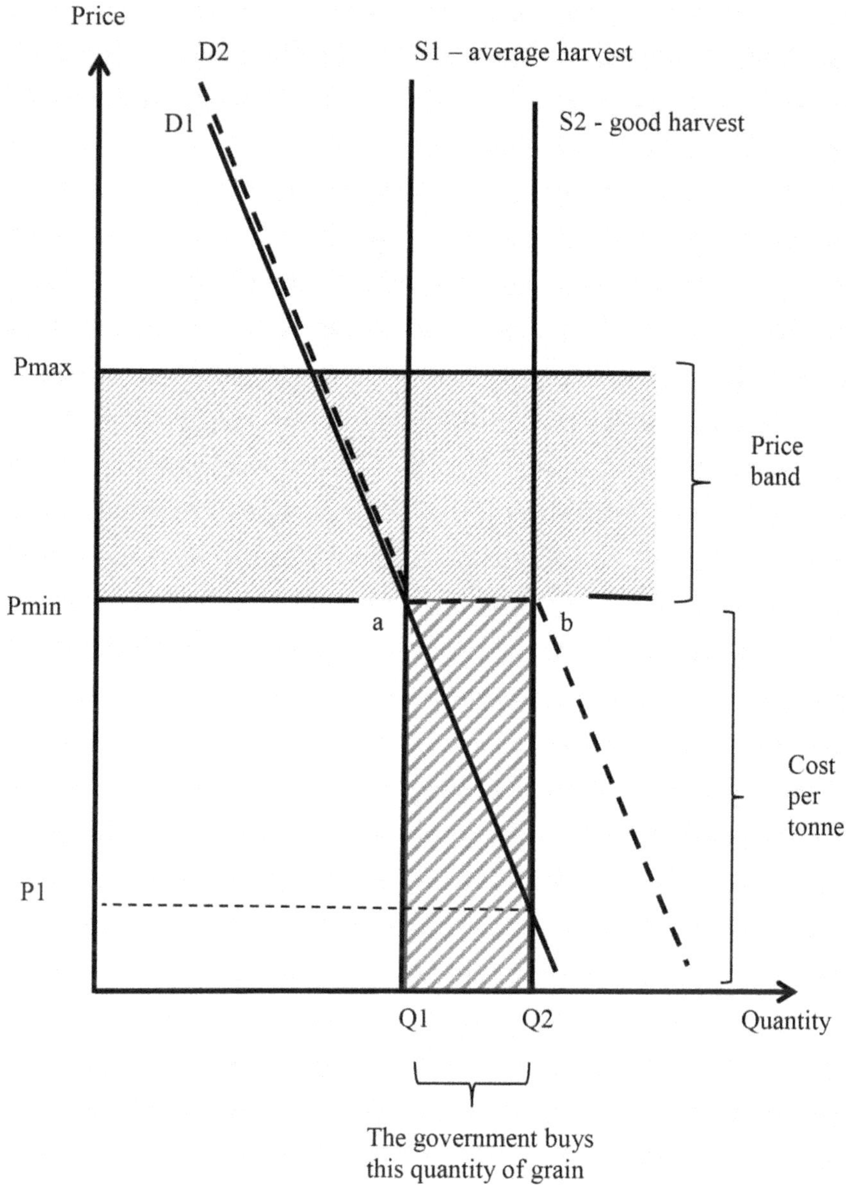

Fig. 9.2. The cost of purchasing grain in a year of good harvest.

In year 5, the domestic equilibrium price again presses against Pmax and the government sells grain to prevent the price exceeding Pmax. The government sells 20 tonnes, bringing the buffer stock down to 40 tonnes. In year 6 the situation repeats itself, with the government now finding itself selling an even greater quantity in order to keep the domestic equilibrium

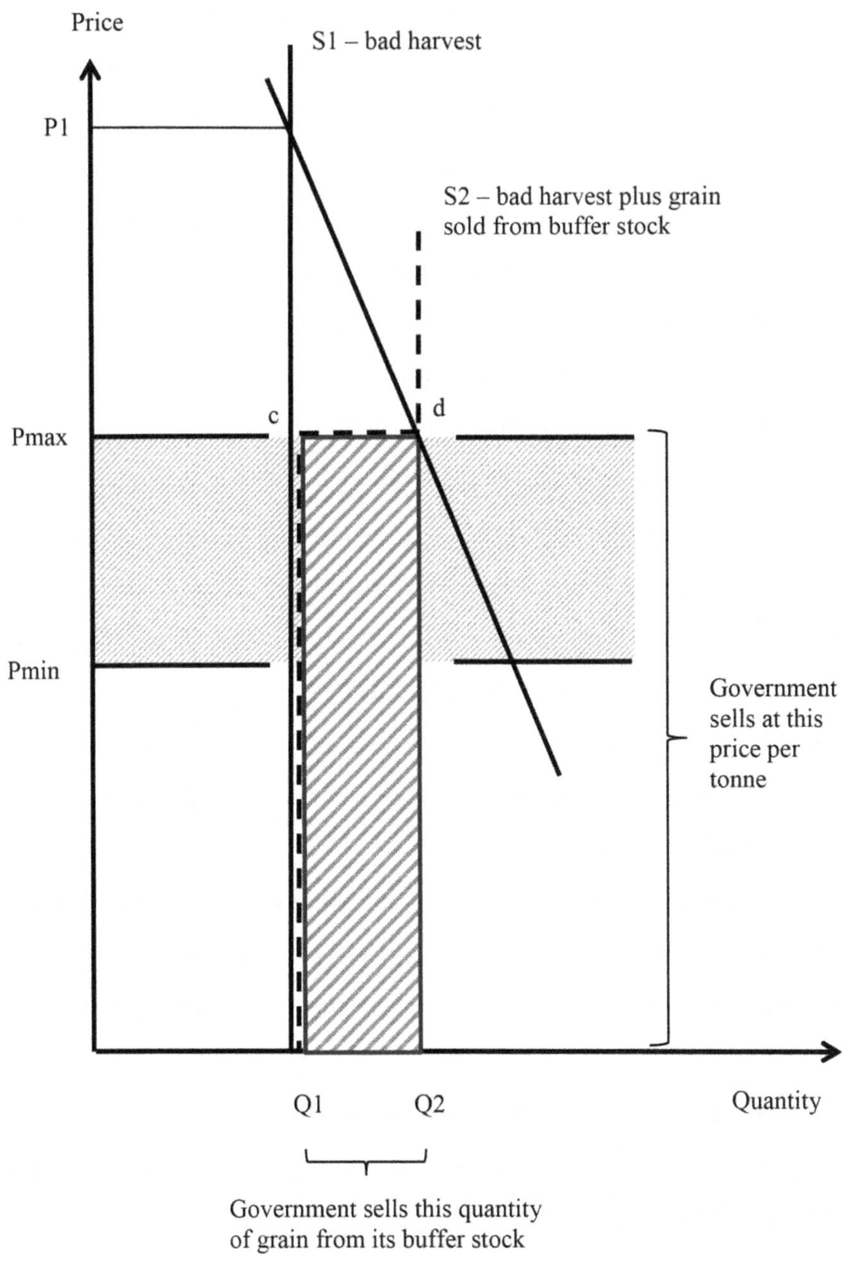

Fig. 9.3. The revenue generated when selling grain from buffer stock.

price within the price band. It has to sell 30 tonnes. The buffer stock is nearly empty and the government is apprehensive that the following year may see a 'stock-out'.

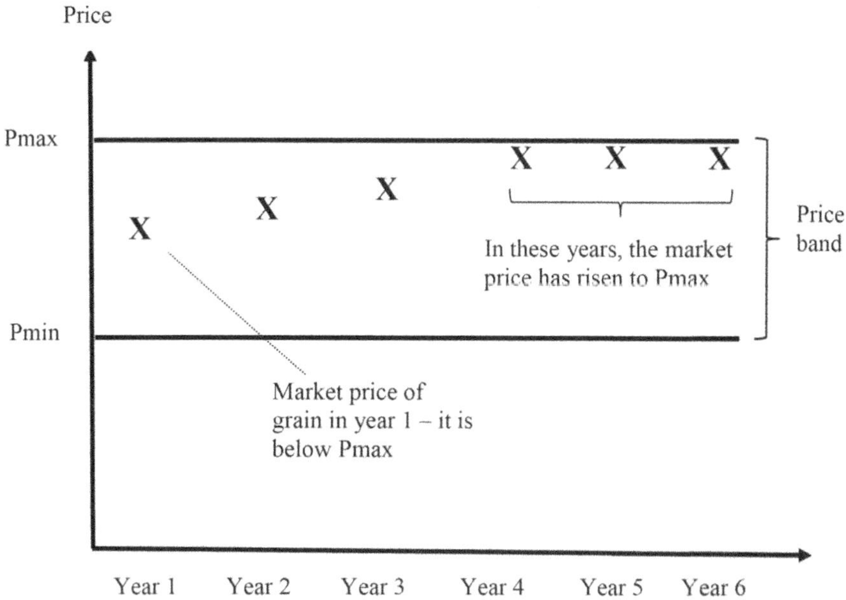

	Year 1	Year 2	Year 3	Year 4	Year 5	Year 6
Quantity of grain sold	No grain is sold	No grain is sold	No grain is sold	10 tonnes is sold	20 tonnes is sold	30 tonnes is sold
Quantity of grain remaining in the buffer stock	70 tonnes	70 tonnes	70 tonnes	60 tonnes	40 tonnes	10 tonnes

Fig. 9.4. Avoiding the depletion of the buffer stock.

What the above information tells us is that the grain sector is evolving, it is not static. But we cannot see the evolution directly. The very actions of the buffer stock do not allow the domestic equilibrium price to reach its 'natural' level. If the buffer stock did not exist, the domestic equilibrium price would rise (or fall) to its natural level and the trend in the domestic equilibrium price, over time, would be clear. All we know is that each year more and more grain has had to be sold to keep the domestic equilibrium price within the price band. In economics, it is said that the long-term equilibrium price is rising. This may be due to successive contractions in the supply of grain or to successive expansions in the demand for grain. Whatever the reason, the government needs to take this long-term trend into account and adjust its price band accordingly. In this particular example, the price band needs to be adjusted upwards.

Of course, the opposite situation may prevail – each year, to prevent the domestic equilibrium price falling below Pmin, more and more grain may need to be purchased and placed in the buffer stock. This will be because the long-term equilibrium price is falling. The price band needs to be adjusted downwards. If not, the buffer stock will soon reach its legal capacity. Then it will not be able to take in any more grain. If that happens, it will not be able to prevent a fall in price in a subsequent year. It is necessary for the government to be ready to adjust the price band to reflect the evolution and trend of the long-term equilibrium price.[5]

9.4.2 The storage period

Humanity has stored grain as food since it started farming some 10,000 years ago. It is not difficult to store grain, providing that the grain is clean and dry before it is placed in the store. Once inside the store, it should be kept cool and aerated. It is important to take measures to control fungi, nematodes and rodents. It is possible, in some climates, to store grain for ten years or more without serious deterioration. But it is best to limit the storage period to one or two years. That way, the grain remains fresh and keeps its monetary value.

9.4.3 Location of the buffer stock

The buffer stock can either be located in a dedicated warehouse or within the commercial pipeline. The grain pipeline is the system whereby commercial grain is harvested by farmers, stored by traders, processed by millers and bakers and eventually delivered as food to the shops and supermarkets. If the buffer stock is stored in a dedicated warehouse the government has to turn it over, at least every two years. This means selling the old grain and replacing it with new grain. The old grain will still be useable but it will have lost a little value because it will no longer be completely fresh. The government will have to buy new grain at a higher price than it receives for the old grain. The government thus bears a cost.

The advantage of embedding the buffer stock within the existing commercial pipeline is that the grain does not lose its value because, as it moves forward through the pipeline, it is constantly replaced by new grain. All the grain, both the buffer stock grain and commercial grain, moves through the pipeline on its way from the farmers to the consumers. The government does not have to turn the grain over and therefore avoids the loss incurred when selling old grain that has been in a dedicated store for a year or two. Figure 9.5 shows the two alternative approaches to locating the buffer stock.

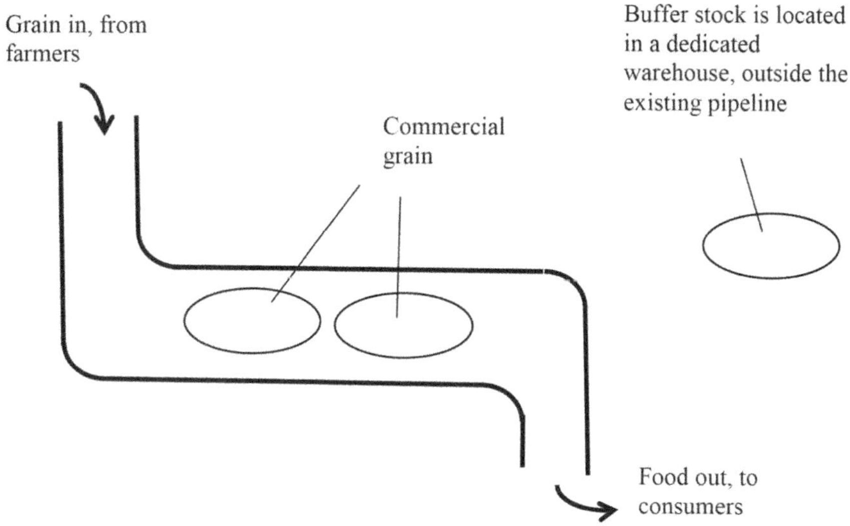

a) the buffer is located outside the existing food pipeline

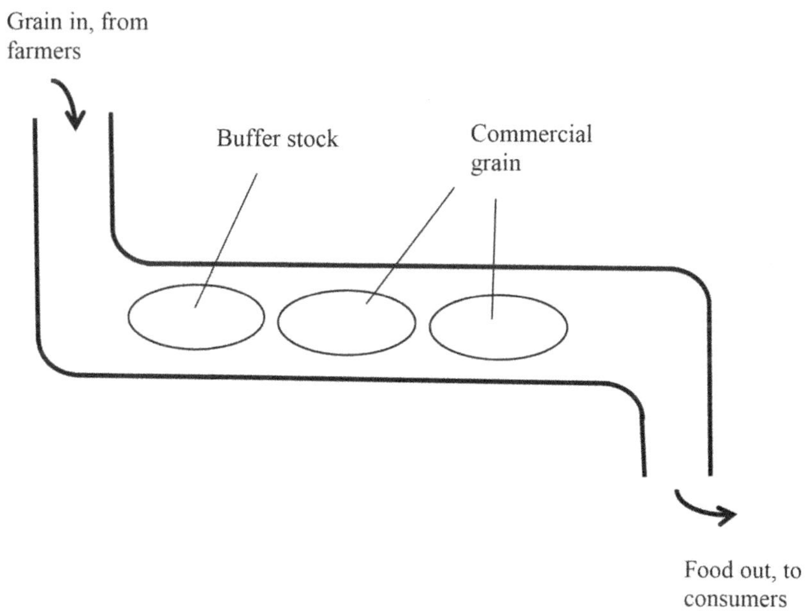

b) the buffer stock is embedded within the existing food pipeline

Fig. 9.5. The location of the buffer stock.

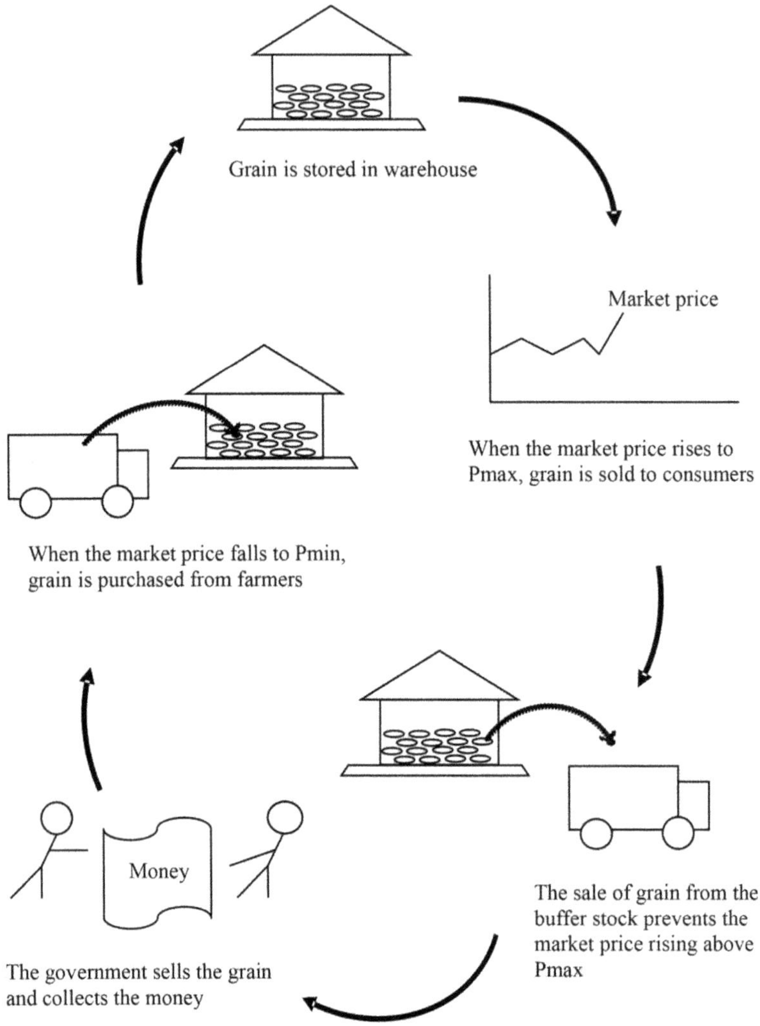

Fig. 9.6. A buffer stock – the cycle of sale and purchase.

9.4.4 The operational cycle

The government follows a cycle of sale and purchase, as shown in Fig. 9.6.

9.4.5 The release of grain from the buffer stock

As explained in Chapter 6, a buffer stock places the government in a position to augment the supply of grain from its farmers onto the market. The government simply has to set up a buffer stock and announce that traders, food processing firms, millers, bakers and all other consumers have the right to purchase grain from the buffer stock at a price of Pmax for as long as there is grain physically available. When the consumers arrive at the government's warehouses, the government sells them grain. The right to purchase grain, of course, cannot be respected if there is a stockout. As long as there is a source of grain at Pmax, a rise in the domestic equilibrium price above this level will be prevented.

In practice, the government may have to be more proactive than simply waiting for consumers to come to its warehouses to purchase grain. This is because the market price may rise above Pmax before consumers start to buy grain from the buffer stock. There may be a time lag between the rise in the price of grain and consumers switching from their traditional source of grain to the buffer stock. Rather than waiting for consumers to come to it, the government may need to proactively release grain from its buffer stock onto the market.

To do this, the government needs to monitor the price of grain on the market. Many governments already collect data on grain and food prices, together with data on the price of other items in the economy. They collect these data for the purpose of economic planning. Monitoring of grain and food prices can take place every week and in every region of the country. The data has to be transmitted to the authority responsible for the management of the buffer stock, usually the Ministry of Agriculture and Food.

In addition to these regular price data, the authority needs to have a clear and precise understanding of the grain economy. This includes estimates and forecasts of the size of the harvest, the quantity of grain stored by private merchants and the volume of imports and exports of grain. With this information, the authority can build up a good knowledge of the factors that influence the evolution of the price of grain.

When it comes to releasing grain for sale on the market, the authority needs to make two decisions: how much grain to release and when to release it. The answers to these questions are provided by the market. The authority itself does not have to provide the answers. Let us first consider how the authority can prevent the market price rising above Pmax. It needs to set a trigger price at a level that is slightly below Pmax. Whenever the market price surpasses the trigger price, this is a signal to the authority that it needs to release a quantity of grain. The second question is how much grain to release. If too little grain is released, it may not be enough to prevent the market price rising above Pmax; if too much grain is released it may cause an unnecessarily large fall in the market price. In practice, the authority can release a relatively small quantity, observe the reaction of the market and, if necessary, release a further quantity. If the authority has good price data and a sound knowledge of the grain economy, it will be able to judge from the reaction of the market what is the appropriate quantity to release each day or week. Figure 9.7 illustrates the procedure.

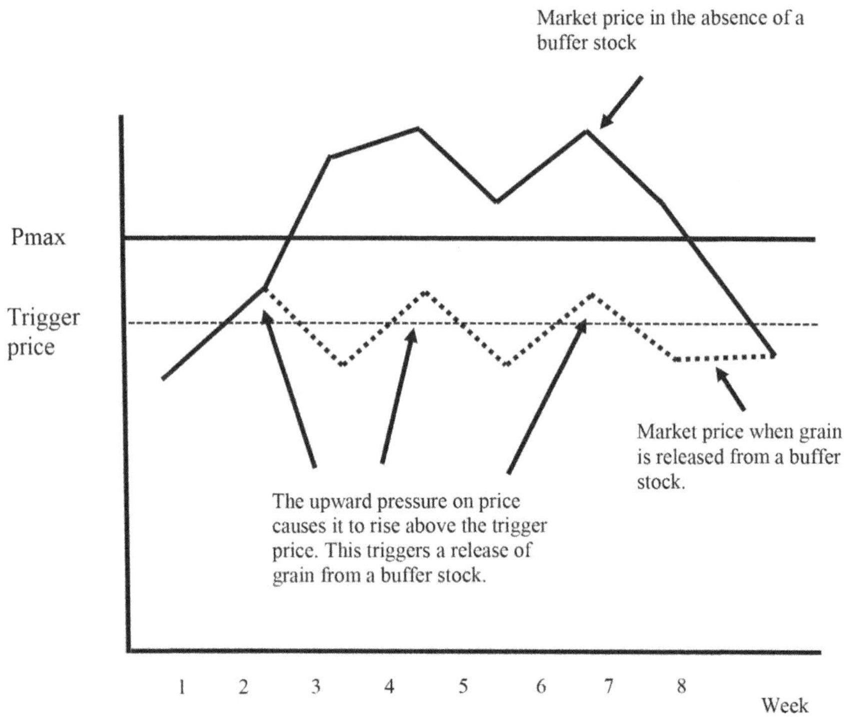

Fig. 9.7. Decisions regarding the timing of the release of grain.

9.4.6 Deciding the quantity to be purchased from the market

When the market price falls below Pmin, the government buys grain from its own farmers. The practical procedure is similar to that of releasing grain, the decision to purchase being triggered when the market price falls below a trigger price and the quantity to be purchased being decided on the basis of market reaction, knowledge and experience.

9.5 Practical Aspects

9.5.1 The limitations of a buffer stock

The capacity of a buffer stock is the maximum number of tonnes of grain that can be stored, and subsequently released. This will be specified in the national law that brings the buffer stock into existence as a legal entity. Other things being equal, the smaller the capacity, the higher the risk that the buffer stock will be depleted (a stockout). Although they can be complicated and expensive, the government may have to resort to rationing grain and subsidizing imports (see 6.5.2 and 6.5.3).

A buffer stock can be replenished only if there is spare capacity to take the new grain. If a buffer stock is physically, and legally, at full capacity, it is not possible to place more grain in it. This is the counterpart to a stockout. The greater the capacity of the buffer stock, the less frequently it will reach its full capacity. If prices risk falling below Pmin and the buffer stock has reached its full capacity, the government may have to subsidize exports or provide farmers with deficiency payments.

9.5.2 A buffer stock is compatible with the market

A market that functions well is a public good. It continuously receives signals from producers regarding their costs, quantities of production and so forth. It also continuously receives signals from consumers regarding what they wish to buy, in what quantities, etc.

The market assimilates all this information and, in turn, transmits it back to both producers and consumers in the form of market prices. On the one hand, market prices tell producers what the consumers wish to buy and at what price. In this sense, prices act as incentives and disincentives to producers regarding the allocation of their factors of production to different activities. On the other hand, market prices tell consumers what the producers are willing to sell and at what price. A well-functioning market is constantly sending messages from consumers to producers and back again, and does this at no charge to society.

Is a buffer stock compatible with a market? Does it not impair or inhibit this process of forming prices and transmitted them, as economic signals, to farmers and consumers?

So long as governments buy and sell grain at the prevailing market price, a buffer stock does not impair the process of price formation or inhibit the transmission of prices to market participants. When a government buys grain from its farmers at the prevailing market price, it acts as any other buyer, whether this be a supermarket wishing to acquire grain to sell to its customers, a trader wishing to export to another country or a miller wishing to buy grain to mill into flour. A government wishing to buy grain to replenish its store, at the going price, is acting just as any other buyer. The efficiency of the market as an information system is not impaired.

Similarly, when a government sells grain from its buffer stock onto the market, at the prevailing market price, it acts like any other seller – whether this be a farmer selling the grain that he has harvested or a trader selling some of his inventory. Again, the efficiency of the market as an information system is unaffected.

In other words, so long as governments buy and sell at the prevailing market prices, buffer stocks are entirely compatible with a well-functioning market. If the government were to buy and sell at prices that are different from the prevailing market prices, then artificial signals would be generated. In that case, either all the farmers would sell to the government or all the consumers would buy from the government. This is the reason why the government needs to act like any other commercial operator – buying and selling at the prevailing market price.

Providing that governments buy and sell at the prices that prevail in the market, the market remains intact as an allocative and distributive system with all the benefits that markets bring to society. The price mechanism continues to operate. Market prices are equilibrium prices. These continue to be determined by the forces of supply and demand. Prices reflect market fundamentals and send signals to producers, traders, private storers and consumers alike. These signals act to maximize economic efficiency.

9.5.3 A buffer stock removes 'noise' and improves the allocation of resources

Buffer stocks can remove the 'noise' from market prices. By so doing, they improve the allocation of resources which, in turn, contributes to a more efficient economy. Random and extraneous events – such as good and bad weather and damage to grain during the growing season from plant pests and diseases – affect the size of the harvest. In some years, the growing conditions are average and a harvest of average size is supplied by the farmers to the market. In other years, the growing conditions are bad, in which case the farmers supply a small harvest to the market. In those years when the growing conditions are exceptionally good, the farmers supply a large harvest to the market. As we have seen from previous chapters, a small harvest causes grain prices to be high while a large harvest causes grain prices to be low. In other words, the price of grain fluctuates from one year to the next due (at least partly) to random, extraneous events such as the vagaries of the weather and the level of plant pests and diseases.

In deciding the area of land to plant to grain in the forthcoming season, farmers need to have some idea of the price that they are likely to receive when they come to sell it. They tend to base their decisions, at least partly, on past prices. Ideally, these should be shorn of the influence of short-run random events. The price would then reflect the longer-term balance between the resource cost of producing grain and market demand. So long as the price band is periodically adjusted to take account of the trend in the long-term equilibrium price, the effect of a buffer stock is to remove some or all of the random element within the market price. By removing this random element from the market price, a buffer stock brings about a better allocation of resources and, in this way, contributes to a more efficient economy.

9.5.4 Price neutrality

In years when the market is tight, the government sells grain onto the market, thus preventing further upward increases in the market price. In years when the market is slack, the government procures grain from the market, thus preventing further downward falls in the market price. The average market price over time will not change. What changes is that the market prices will be more stable – the variance of price will be less.

9.5.5 Does a buffer stock invite speculation?

No government can undertake to keep the grain price within the price band under all possible circumstances. This would place an intolerable burden on the government and invite speculation against it. The government is under a legal obligation to prevent the equilibrium price rising above the price band *only* for as long as there is grain in the buffer stock that can be sold. If there is a stockout, the government is not in a position to guarantee that the price of grain will remain within the price band. Similarly, the government is under a legal obligation to prevent the equilibrium price from falling below the price band *only* for as long as there is spare capacity in the buffer stock for more grain to be taken in. By limiting itself only to what is clearly feasible, the government does not unduly bind itself. This means that there is no scope for speculators to 'bet' against the government.

9.5.6 Will not buffer stocks end up as unwanted farm surpluses?

What is to prevent a buffer stock becoming ever larger and end up being a depository of farm surpluses? How will a government be able to avoid repeating the difficulties that befell the USA and the EU with their stocks of surplus commodities?

In the case of the surplus stocks accumulated by the USA and the EU during the 1960s–1990s, these were open-ended and legal obligations to buy unlimited quantities of grain. The floor price, Pmin, was deliberately set at a level above the domestic equilibrium price, in order to raise farmers' incomes on a permanent basis. In contrast, a buffer stock, as discussed in this book, is a specified quantity of grain held by a government for the purpose of stabilizing the country's grain market (not for permanently increasing farmers' incomes). The floor price, Pmin, is not set above the domestic equilibrium price – it is set below it. Moreover, the buffer stock is of a fixed maximum capacity (i.e. *x* thousand tonnes). This maximum capacity is specified in the relevant legal act.[6] If the government buys in excess of the maximum capacity of the buffer stock, it will be in breach of the law. There is thus a legal safeguard against a buffer stock becoming a repeat of the saga of the structural surpluses of the USA and the EU.

9.6 Legal Aspects – The Law of the World Trade Organization

The issue of stabilization of agricultural prices has raised a number of debates among the members of the World Trade Organization. It is important that buffer stocks are set up and operated in accordance with international law. In this section, we discuss the provisions of law of the WTO as it currently stands.

The provisions of the WTO that concern agriculture are contained in the Agreement on Agriculture (AoA). There are no provisions that prevent or prohibit a government from setting up a buffer stock. Many members of the WTO have buffer stocks, and their establishment has not been called into question or contested. However, the AoA does have something to say about the purpose of a buffer stock and the manner in which it is operated. To understand this, we need to appreciate the ethos and *raison d'être* of the AoA itself.

Farming in many countries had been supported by governments and protected from international competition since World War II. This was felt to be detrimental from the point of view of global economic progress. It was necessary to open up agricultural trade. It was also necessary to place a limit on the level of support that governments could grant their farmers. To achieve this latter goal, governments agreed, first, to calculate the level of support in terms of its value (in currency units) and, secondly, to reduce it by a certain percentage. For instance, if a country calculated that its level of support to its farmers was $3 billion per year and if it had to reduce it by 33%, the country would be obliged to reduce its support in the future to $2 billion per year. This could have huge implications for how a government supported its farming community. Programmes which had formerly been considered a standard part of agricultural policy, might have to redesigned or even stopped.

Support to farmers comes in many different shapes and sizes. Some forms are clearly a constraint on international trade (such as taxes on imports and subsidies on exports). Other forms of support are less constraining and may have little impact on international trade (such as money paid to farmers to stop soil erosion). The negotiators realized that if governments were going to agree to reduce the level of support they provided to their farmers, it would be necessary to distinguish between those forms of support that distorted international trade and those forms that did not. Accordingly, the AoA gives a list of those forms of support to agriculture that are considered not to distort agricultural trade. This list constitutes Annex 2 of the AoA.

The purpose of Annex 2 is to identify those forms of support that do not distort international agricultural trade. In the jargon of the WTO, these are 'green box' measures. Such measures are defined, in the first paragraph of Annex 2, as measures which 'have no, or at most minimal, trade-distorting effects or effects on production'. Because they are considered not to distort international trade, green box measures are exempt from the obligation to reduce the level of support to farmers.

Returning to the matter of a buffer stock, it is important to know whether this is, or is not, a green box measure. Paragraph 3 of Annex 2 is the relevant provision and reads as follows (World Trade Organization, 2016):

> Paragraph 3. Public stockholding for food security purposes:
>
> Expenditures (or revenue foregone) in relation to the accumulation and holding of stocks of products which form an integral part of a food security programme identified in national legislation [are a green box measure]. This may include government aid to private storage of products as part of such a programme.
>
> The volume and accumulation of such stocks shall correspond to predetermined targets related solely to food security. The process of stock accumulation and disposal shall be financially transparent. Food purchases by the government shall be made at current market prices and sales from food security stocks shall be made at no less than the current domestic market price for the product and quality in question.

In ordinary parlance, and with the caveat that only judges can provide definitive interpretations of legal texts, paragraph 3 can be unofficially translated as: a buffer stock is a green box measure, providing its purpose is food security, that the financial accounts are clear and available, that the government buys grain to replenish the buffer stock at the current market price and that the government sells grain from the buffer stock at the current market price.

As explained in Section 6.4 (the economics of a buffer stock), for a buffer stock to work, the government needs to buy and sell grain at the market price – in other words in accordance with paragraph 3 of Annex 2. So long as a country buys grain for the buffer stock and sells grain from the buffer stock at market prices, the economic cost of the buffer stock is exempt from the country's obligation to reduce its level of support to its farmers. This is because buying and selling at the market price does not distort international trade.[7]

9.7 Private or Public Storage?

If markets were working perfectly and there was perfect information at all times and in all countries, there would be neither food shortages nor food surpluses. Entrepreneurs would operate private stores of grain, adding grain to them in times of low prices and releasing grain in times of high prices. These entrepreneurs would operate their stores of grain as a business. As with any other business, the objective would be to make a living. There would be no need for the government to operate a public buffer stock. The price mechanism of the market would even out the high and low prices. A problem of price volatility would not arise. The fact that price volatility does occur – and causes distress for consumers and farmers alike – indicates that, for some reason or other, entrepreneurs are not operating private grain stocks, or at least not at the scale that is required to prevent the levels of price volatility to which the world has recently been subjected. In other words, private storage of grain, on a multi-annual basis (as opposed to a seasonal basis, i.e. from one harvest until the next harvest), is inadequate relative to the quantity that is required to stabilize grain prices from one year to the next.

Among the reasons for inadequate private storage of grain are the following:

1. Private storage is inherently risky: it may be several, indeed many, years until prices rise to the level that gives the storer a profit on the grain in his silo. Every year that the grain is in storage, the storer incurs costs. Professor Holbroke Working (1953, p. 561) of Stanford University observed that 'The warehousing of surplus commodity stocks is a very uncertain and hazardous business when based on trying to judge when the price is favourable for storage...'
2. When the storer does come to sell his grain, he may be accused of 'hoarding food solely to profiteer from the misfortune of those who are poor and hungry'. Storers have been beaten up, arrested and in periods of extreme food shortages, such asduring a war, have even been imprisoned and threatened with capital punishment.
3. At times of high food prices, governments tend to intervene in the market. They attempt to calm the market and prevent the price of food becoming unaffordable. One way of doing this is to impose a ceiling price and forbid traders from selling grain at a price higher than the ceiling price. This can wipe out the potential profit from storing grain in the first place. Governments may also confiscate grain stored by private entrepreneurs. The reader may be surprised to learn that that bastion of free enterprise – the United States government – confiscated some of the oil stocks of private oil companies in 1973 when the country was subjected to an oil embargo (Williams and Wright, 1991, p. 418).

The uncertain profits and the risks associated with private grain storage probably explain why the quantity of stocks held by entrepreneurs, on a multi-annual basis, is not sufficient to even out the fluctuations in the quantity of grain supplied to the market. The private storage of grain is replete with risks. It may bring a profit, but it may bring a loss. The tendency for private stocks to be less than the quantity that a country needs in a food emergency means that it is the government that has to take the responsibility for moderating the volatility of the price of grain. Price moderation should be regarded as a public good, since it benefits everybody. Only the government is in a position to provide public goods.

9.8 National or Regional (Multi-country) Buffer Stocks?

Countries can set up their own national buffer stock for their own use. However, they can save money if they join together with other countries and share a buffer stock at the regional level. Buffer stocks are an insurance policy – the more members, the less risk is incurred by each member and the lower the individual premium. The International Food Policy Research Institute (1986) calculated that a regional grain reserve for the countries of southern Africa would require 40% less grain than the sum total of reserves.[8] Regional buffer stocks can replace national buffer stocks. Alternatively, a regional stock can exist alongside a national stock and complement it.

Most countries in the world belong to one or other regional economic integration organization. These organizations are well placed to set up and manage a regional buffer stock on behalf of their members. The member governments are responsible for the major decisions. The organization remains under the overall control and supervision of its member governments.

This is the principle that most European countries have chosen in order to work together for the common good. It has proved successful – since the EU (EEC) was formed in the 1950s, the states have not engaged in war and the people have gradually become better off. Compared to previous centuries, Europe is now very prosperous and one of the reasons for its prosperity is its integrated economy, made possible by the EU. A regional buffer stock is perhaps a way for other regions of the world to start on the same path towards peace and prosperity.

9.9 Conclusion

This chapter has argued that grain buffer stocks can be effective in moderating the volatility of the price of grain on national markets. They are compatible with the market economy and improve economic efficiency. They have a neutral effect on price, neither increasing nor decreasing the average price throughout the year. They do not invite speculation and do not end up as unwanted surpluses of grain. Buffer stocks are compatible with the rules of the World Trade Organization. Given these positive attributes, why has the idea of buffer stocks become somewhat unfashionable among some governments and economists? The next chapter addresses this question.

Notes

[1] We can note, in passing, that, in the academic literature, there are rather few in-depth studies of buffer stocks. There is a paucity of objective, quantitative information on issues such as running costs and practical performance.

[2] China joined the World Trade Organization in 2001.

[3] Huang *et al.* (2013) have provided a full account of China's food reserve in a paper for the United Nations University. They have written: 'The government of

China, in responding to the crisis, chose a specific set of policy measures that included the release of the government's grain reserves in the beginning of the crisis, long-term future/forward contracts with trading firms in exporting countries, provision of subsidies and insurance to producers, cancellation of support for storage and transport of export grains, increased subsidies on grain production and input, and enhanced social protection for urban consumers. Similarly to many other countries, however, China also used wider measures at the border to protect domestic prices from international food price fluctuations. Furthermore, China adjusted its long-term development strategies on biofuel development and strengthened the commitment to invest in agriculture.' The authors explain that the grain market had already been liberalized: 'China's agricultural markets liberalized gradually. Farmers generally decide what crops they want to plant. Furthermore, they purchase all the necessary chemical fertilizers from private vendors on their own with no involvement of local officials. The majority of grain, oilseed and fibre crop sales, as well as virtually all purchases of horticultural and livestock products are handled through millions of small private traders, a departure from the 1980s when farm output was purchased by the government's procurement agencies. The presence of millions of small farm-gate and wholesale traders competing with virtually no regulation implies that the markets in China have become integrated and efficient.' (p. 4)

[4] The benefits that China draws from its buffer stock have been examined in a number of papers, in addition to Huang *et al.* (2013). For instance, Jensen and Miller (2008) report that, in Hunan and Gansu provinces, the global food crisis of 2007/8 did not cause a decline in nutritional status. See also Harkness (2011) for an account of how China coped when less than half an inch of rainfall in Shandong Province, one of the country's main grain-growing areas.

[5] Ritson (1977, p. 341) points out how the government can resolve this difficulty of adjusting the price band: 'A buffer stock must aim to stabilise around a trend price, the difficulty of doing so being to identify the trend when the operation of the stock is itself determining the level of ... prices for the commodity. One possible way out of this impasse would be to agree to limit the size of the stock to whatever level thought necessary to even out the short-term supply fluctuations, and to apply a formula to alter the [price band] as the size of the stock approaches either zero or its upper limit.'

[6] The maximum capacity must be adequate for the purpose of stabilizing the price. If it is underestimated, the government will not be able to purchase sufficient grain in years of good harvest to prevent the market price falling below Pmin. Likewise, in years of bad harvest, the government will not be able to sell sufficient grain to prevent the price rising above Pmax.

[7] Having said this, food reserves have proved controversial within the WTO with a dispute that started in 2002 and which remains unresolved. The dispute does not concern food reserves that operate at market prices (as this book proposes); rather, it concerns reserves that operate at 'administered prices', that is prices which are set by the government. If the government purchases grain from its farmers for its reserve at prices that are higher than the market price, this constitutes price support. The grain reserve cannot be considered to be a green box measure. Expenditures regarding the reserve have to be included in the calculation of the total level of support provided by the government to its farmers and, consequently, may be subject to a limit. This has proved a contentious issue

with India and a number of other countries. If, for instance, India was obliged to reduce the economic cost of its food reserve, it could have a very substantial effect on its farm support programmes and its level of food security. The issue has been repeatedly discussed at technical and ministerial meetings within the WTO, but at the time of writing (2020) it remains unresolved.

[8] The study was written by Professor Ulrich Koester of Kiel University in Germany. He has written: 'A regional stockpiling system for grains is a strategy that would not require the sacrifice of national autonomy but would allow for increased food security. Hence, this alternative is investigated in detail. Based on past fluctuations in cereal production and import prices, the amount of stocks needed for each country to stabilize cereal consumption is calculated, and is compared to the stocks required by the same countries cooperating regionally. Results show that regional stocks could be about 41% less than the sum or national stocks without cooperation.'

References

Dawe, D. and Timmer, P.C. (2012) Why stable food prices are a good thing: lessons from stabilizing rice prices in Asia. *Global Food Security* 1(2), 127–133. DOI: 10.1016/j.gfs.2012.09.001.

Harkness, J. (2011) *Food Security and National Security: Learning from China's Approach to Managing Its Wheat Supplies.* Institute for Agriculture and Trade Policy, Minneapolis, Minnesota. Available at: http://www.iatp.org/files/2011_02_28_FoodSecNatSec-JH.pdf

Huang, J., Yang, J. and Rozelle, S. (2013) *The political economy of food pricing policy in China,* WIDER Working Paper 38. Unu-Wider, Helsinki.

International Food Policy Research Institute (1986) *Regional Cooperation to Improve Food Security in Southern and Eastern African Countries,* research report 53. Ulrich Koester, Washington.

Jayne, T.S., Zulu, B. and Nijhoff, J.J. (2006) Stabilizing food markets in eastern and southern Africa. *Food Policy* 31(4), 328–341. DOI: 10.1016/j.foodpol.2006.03.008.

Jensen, R.T. and Miller, N.H. (2008) The impact of food price increases on caloric intake in China. *Agricultural Economics* 39(1), 465–476. DOI: 10.1111/j.1574-0862.2008.00352.x.

Ritson, C. (1977) *Agricultural Economics: Principles and Policy.* Crosby Lockwood Staples, UK.

Williams, J.C. and Wright, B.D. (1991) *Storage and Commodity Markets.* Cambridge University Press, Cambridge, UK.

Working, H. (1953) Hedging reconsidered. *Journal of Farm Economics* 35(4), 544–561. DOI: 10.2307/1233368.

World Trade Organization (2016) *The WTO Agreements Series: Agriculture,* 3rd edn. Geneva. Available at: https://www.wto.org/english/res_e/booksp_e/agric_agreement_series_3_e.pdf

10 Attitudes towards Price Stabilization and Buffer Stocks

10.1 Introduction

Attitudes towards the stabilization of commodity prices and the role of buffer stocks have a tendency to wax and wane over the decades. The purpose of this chapter is to trace, in a rather rough and ready manner, how attitudes have changed over the last 100 years.

10.2 The International Wheat Agreement of the 1920s

In the 1920s, the major governments of the world sought to stabilize wheat prices on the world market. They negotiated an International Wheat Agreement. This was the first international attempt to bring some calm and predictability to a commodity market. The agreement, which is described in the Appendix, was partially successful. It lapsed in the 1960s.

10.3 International Commodity Agreements

After World War II, various governments endeavoured to stabilize the world price of a number of commodities by means of 'international commodity agreements'. Agreements were reached for coffee, cacao, rubber, sugar and tin. Like that for wheat, the agreements have not survived. Table 10.1 shows their lifetimes and the reasons for their eventual collapse or suspension.

The most spectacular collapse was that of the Tin Agreement in 1985. It collapsed because the buffer stock manager had acquired a very large holding of tin at a time when demand for the commodity was falling. The collapse of this agreement led many economists to believe that it is not possible to control the price of a commodity and that any attempt to do so will inevitably fail. In this vein, *The Economist* proclaimed that 'the first lesson is that the road down which the tin men have clattered – that of artificial control of a market – leads to ruin' (*The Economist*, 1985).

Table 10.1. International Commodity Agreements.

Commodity	Date	Breakdown or lapse
Tin	1954	Collapsed in 1985 – was attempting to hold the price at too high a level with insufficient finance
Sugar	1954	Lapsed in 1983 – due to ineffectiveness
Coffee	1962	Suspended in 1989 – lost the support of consumer countries
Cocoa	1972	Suspended in 1988 – due to ineffectiveness
Rubber	1980	Terminated in 1999 – due to internal divisions (downward revision of the floor price was controversial)

Professor Christopher Gilbert of the University of Trento has studied the international commodity agreements. He has explored the reasons that led to their demise. Referring to *The Economist*'s conclusion that any attempt to control the price of a commodity leads to ruin, Gilbert has written: 'Comments of this sort ignore the fact that the tin agreement was successful in controlling the tin price for 25 years; they ignore the success of OPEC in controlling oil prices over the decade following 1974; they ignore the remarkable ability of de Beers to control the world diamond market and they suggest an over simple explanation for the more general lapse of the international commodity agreements' (Gilbert, 1996, p. 1).

Although the tin agreement collapsed in bankruptcy, none of the other agreements followed the same road of financial ruin. They did not collapse for financial reasons. Rather, they lapsed or were voluntarily closed by their member countries. In sugar, this was because of adverse market conditions, which made any attempt at stabilization impractical. In cocoa, there was never sufficient support for stabilization for the authority to have the funds to intervene effectively in the market. In the coffee market, stabilization was effective both in raising prices and containing their variability, but there was disagreement over the division of the benefits between countries and because the effects of high prices did not often reach the coffee farmers. In contrast, the rubber authority intervened on such a low level as to cause little enthusiasm for the agreement among its member countries (Gilbert, 1996).

10.4 A Loss of Interest in the 1980s

The lack of evident success of these international commodity agreements led, during the 1980s, to price stabilization losing its allure and its appeal. It was no longer a prominent item on the agenda of the UN. Economists seldom recognized its importance. For example, Amartya Sen, winner of the Nobel Prize for economics, chose not to include the stabilization of commodity prices as a key element of his 'support-led security' measures that he advocated at that time for developing countries (Newbery and Stiglitz, 1989). But also important were three other developments of the 1980s: political ideology, the experience of the structural agricultural surpluses of the USA and EU, and the doubts expressed by some academics.

10.4.1 The neoliberal ideology of the last 40 years

Before the 1980s, there was a general belief – at least in the non-Communist western countries – that the best way to run an economy was by a mix of market forces and state intervention. Broadly speaking, this is shorthand for a social market economy. Most western countries were run on these lines. Some decisions were left to the private sector, to business and to the market, but other decisions – those of a more strategic nature – were taken by the government, acting on behalf of the state.

After the 1980s, beliefs changed. The writings of Milton Friedman (Friedman, 1962) of the University of Chicago became popular. His view was that proactive governments were inimical to economic efficiency and individual freedom. He argued that it was better for governments not to intervene in the economy. As far as possible, governments should let the economy run itself. Governments should leave decisions to the market, i.e. to entrepreneurs, businessmen and -women, bankers, farmers, traders and so forth. The belief was that the interaction of their individual private decisions would produce an outcome which was best for the economy. Adam Smith's 'hidden hand' would be allowed to operate.

US President Reagan and UK Prime Minister Thatcher adopted Friedman's views. They promulgated his doctrine throughout much of the western world. The result was that free markets and free trade became fashionable. They were the new orthodoxy. State intervention as a general principle was *démodé*. Schemes to stabilize the prices of commodities were not compatible with the new orthodoxy. They did not find favour with western politicians and fell out of favour with the economics profession. The belief that it was a good idea for governments to stabilize the prices of commodities was no longer respectable.

10.4.2 The surplus stocks of the USA and EU

Another reason for scepticism towards grain buffer stocks was the structural agricultural surpluses that arose in the USA and the EU after World War II. We have already alluded to these in previous chapters. These surplus stocks arose because the USA and the EU decided to support farm prices at levels that were often higher than the level of the domestic equilibrium price. The reason that the governments decided to support farm prices was because they wanted to support the incomes of farmers – to avoid the standard of living of farmers falling behind that of the nation as a whole. The high farm prices induced farmers to produce prodigiously. Soon farmers were producing more food than their citizens could consume. This required the governments to buy up the surpluses, lest the market price fell and nullified the goal of price support. The governments bought up the surplus grain, milk and meat and kept it in storage. The controversies provoked by these surplus stocks have become part of the folklore of international agriculture and have entered the consciousness of the economics profession. A suggestion to set up a strategic food buffer stock is sometimes a reminder of these large and unwanted surplus stocks. The issue hung around the necks of the USA and EU until policy was changed in the 1990s. There is naturally a concern lest a food buffer stock repeats the debacle of unwanted surplus stocks.

The point is that a food buffer stock is a way of preventing food prices rising too high and falling too low. The purpose is not to permanently raise farm prices above their equilibrium level. Governments need to make this clear in the legislation setting up a buffer stock and specify a maximum size of stock that can be held.

10.4.3 Academic Studies

Economists have long addressed the question of whether countries benefit when commodity prices are stabilized. Their approach is to calculate the 'economic surplus' that arises when a) there is a stabilization programme in place and b) when there is no such programme in place.

The economic surplus has two elements: consumer surplus and producer surplus. Let us suppose that the market price of a loaf of bread is $3. This is the price that all consumers pay. However, there may be some consumers who are willing to pay more, say $5. If so, these consumers gain a 'surplus' of $2 on every loaf bought. They obtain for $3 an item which to them is actually worth $5. Consumer surplus is the sum total of all the individual surpluses of all consumers.

Producer surplus is an analogous concept. Economists sometimes call it 'economic rent'. Suppose that the price of wheat is $100 per tonne. This is the price that all farmers receive. A farmer's economic rent is the excess of $100 over the cost of producing each tonne sold. Producer surplus is the sum total of all individual surpluses of all farmers.

To derive consumer and producer surplus, economists use supply and demand models. The consumer surplus is the area above the price line and below the demand curve, while the producer surplus is the area below the price line and above the supply curve (the supply curve reflects the cost of producing each tonne that is sold). However, the shape and position of the supply and demand curves depend on the behaviour of consumers and producers, behaviour which may alter as circumstances change (such as whether food is plentiful or scarce, the level of risk that farmers are willing to take and so forth). The economist's estimates of consumer and producer surplus depend on the assumptions that underlie the model of supply and demand.

Even if the economist were able to estimate supply and demand correctly, a second consideration arises that concerns time periods. If a comparison is made between prices fluctuating between a maximum and minimum level and a policy which stabilizes prices half way between the two, then the result is that both consumer and producer surplus is higher with fluctuating, compared to stable prices. The reason is that consumers 'play the market', buying more at low prices and less at high prices.

When consumers play the market over a series of years during which prices are unstable, the average consumer surplus is higher than the average consumer surplus during a series of years of stable prices. It would appear from this result that consumers are better off when prices fluctuate than when they are stable.

Similarly, farmers can play the market by expanding production when prices are high and contracting production when prices are low. This results in a higher average 'economic rent' than when prices are stable. Again, it would appear that farmers are better off with fluctuating, rather than stable prices.

However, averages can be deceptive because they obscure the situation in individual years. For consumers it is of little value to know that next year, if the price of wheat were to fall they would be able to eat more (and benefit from a larger consumer surplus) if, this year, there is a famine. Similarly, farmers are unlikely to be able to exploit the benefits of higher prices in a future year (and benefit from a larger rent) if this year's low prices are causing them to go out of business.

Nevertheless, some economists have used the concept of economic surplus to draw conclusions regarding the wisdom, or otherwise, of price stabilization. Most notable are the studies by Newbery and Stiglitz (1981) and Williams and Wright (1991). The former (p. 23) questioned 'the desirability of price stabilization schemes both from the point of view of the producer and of the consumer'. The latter study (p. 409) was similarly sceptical, arguing that price-band schemes are not beneficial and 'the mystery is why they have so often been advocated by economists'.

Furthermore, the calculus of the economist omits that which is unquantifiable. For example, it is impossible to put a number on the human toll that arises when families cannot feed their children. There are many such human and political effects of price volatility, as already described in Chapter 5, on which it is impossible to put a number. For these reasons (the assumptions used, the use of averages and the omission of the unquantifiable), economic assessments which appear to show that consumers and farmers are better off with unstable prices need to be treated with circumspection.

10.5 Current Attitudes

10.5.1 A view that buffer stocks are expensive

There is a pervasive view that grain buffer stocks are expensive. For instance, the consortium of international organizations led by the OECD in its report of 2011 to the G20 wrote: 'Attempting to stabilise prices using buffer stocks is potentially very costly. Stabilising world prices around a level either lower or higher than that determined by market fundamentals requires significant resources' (OECD and FAO, 2011).

We would agree with this assertion. It is, indeed, very costly to try to stabilize the price of a commodity *around a level that is either higher or lower than the market equilibrium price.* We would not suggest doing this. What we are suggesting in this book is different: to influence equilibrium prices through the mechanisms of supply and demand, thereby moderating price volatility around its long-term equilibrium trend. We are not suggesting going against the long-run equilibrium trend.

Professors Ray and Schaffer of the University of Tennessee point out that buffer stocks are a lot cheaper than the alternatives. In addressing the claim of 'buffer stocks are too expensive', they respond:

> Compared to what? Were buffer stocks ... more expensive than the billions the USA spent in Emergency Payments between 1988 and 2001? Were they more expensive than paying subsidies on every bushel of production rather than storage payments on a portion of production put into a reserve? Were reserves more expensive than 16 years of direct payments? The answer is categorically no. (Ray and Schaffer, 2012)

10.5.2 A reappraisal of the belief that markets can solve all economic problems

An uncritical belief in the ideology of free markets and free trade has recently come under fire. For instance, the then Prime Minister of the UK, Theresa May, stated in her speech to the annual conference of her own Conservative Party in Birmingham in October 2016:[1]

> Where many see government as a problem, I want to show it can be part of the solution. While government does not have all the answers, government can and should be a force for good; that the state exists to provide what individual people, communities and markets cannot; and that we should employ the power of government for the good of the people.

The proponents of grain buffer stocks argue that ensuring that the price of food is always affordable is a task that cannot be left to the private sector. As we argued in the previous chapter, it is not reasonable to expect private farmers and traders to store sufficient grain to tide the whole of society over a year – maybe several years – of poor harvests. The incentives for private entrepreneurs to conduct multi-year storage are simply not strong and robust enough for them to be sure of making a profit. It is a task that falls to governments.

10.5.3 A recognition of the role of buffer stocks

Against this background of scepticism, the world food crises of 2007/8 and 2011, provoked a reappraisal of the potential role of national buffer stocks.[2] For instance, the Chief Economist of the World Bank, Justin Lin, was of the view that buffer stocks could be useful (World Bank, 2008):

> If left to market forces alone, grain production and price will display large volatility, exposing a country to undesirable social and political consequences. A possible remedy is for a government to create a public buffer stock so as to stabilize the market and reduce price volatility … the costs of holding a public grain buffer stock can be viewed as an expenditure for a public good – it mitigates the risks of social unrest and political instability.

Four years later, the World Bank had conducted a review and concluded (World Bank, 2012): 'Public food grain stocks can be used as one of several policy instruments to protect vulnerable people from food price spikes'.

10.6 Conclusion

Commenting on the world food crisis of 2007/8, Peter Timmer of Harvard University observed that 'there has not been any serious attempt to devise a way of stabilising commodity prices at the global level over the last 30 years' (Timmer and Jayne, 2011, minute 51). He continued:

> I wish we [the economics profession] would spend a whole lot more attention on the question of transparency of national food buffer stocks – rather than saying 'don't do it', the question is really how do you make national buffer stocks more efficient?

The reappraisal of the role of buffer stocks is undoubtedly a positive development.

Notes

[1] Speech by Theresa May, UK Prime Minister, at the Conservative Party conference, 5 October 2016; downloaded from BBC report: http://www.bbc.com/news/uk-politics-37556019 (accessed 27 April 2020).

[2] The International Food Policy Research Institute (IFPRI) of Washington went further than proposing national buffer stocks to stabilize prices on national markets. It proposed stabilizing grain prices on the world market by means not of a world physical reserve of grain but by means of a world virtual reserve of money (Von Braun and Torrero, 2009). There would be a fund that could be called upon

in the event that a speculative attack caused the world price of grain to increase. The proposal is based on the premise that speculation in futures contracts causes the price of futures to rise and that this, in turn, causes the spot price to rise. However, as discussed in Section 3.6, there is no consensus that speculation in futures contracts is the cause of volatility of spot prices. There is evidence in support of this thesis but it is not conclusive. For instance, Wright (2009, p. 31) argues against IFPRI's contention that speculation in futures contracts causes price volatility.

The fundamental idea behind the IFPRI proposal is to sell futures contracts, thereby moderating the increase in price brought about by the effect of speculators who are buying futures contracts. The price of futures contracts, it is argued, will no longer rise. This, IFPRI argues, will avoid price volatility. However, for this to work, the fund would have to adopt 'naked short positions'.

Naked short positions arise when a speculator takes out a short position but has no physical grain in store, or growing in the field, to cover it. When it comes to settlement of the contract, the speculator has two choices: he can deliver the physical grain (or money equivalent) or he can offset the short contracts by buying contracts (going long). This is plausible if prices have declined over the life of the futures contract. But a danger arises if prices have risen – this means that the fund would find itself in difficulties. It will end up having sold futures contracts for less than it has to pay in order to either honour them by delivering physical grain, or by having to offset the contracts by going long. For this reason, naked short positions should be avoided. If not, the fund could find itself losing all its money and ending up bankrupt.

References

Friedman, M. (1962) *Capitalism and Freedom.* Chicago University Press, Chicago, Illinois.

Gilbert, C.L. (1996) International commodity agreements: an obituary notice. *World Development* 24(1), 1–19. DOI: 10.1016/0305-750X(95)00121-R.

Newbery, D. and Stiglitz, J. (1981) *The Theory of Commodity Price Stabilization: A Study in the Economics of Risk.* Clarendon Press, Oxford.

Newbery, D. and Stiglitz, J. (1989) *Hunger and Public Action.* Clarendon Press, Oxford, UK.

OECD and FAO (2011) *Price Volatility in Food and Agricultural Markets: Policy Responses.* Report to the G20, co-ordinated by OECD and FAO. 2 June.

Ray, D.E. and Schaffer, H.D. (2012) *Grain Reserves: A Much Needed Discussion, though Some See it as Off-the-table,* Policy Pennings no. 635. Agricultural Policy Analysis Center, University of Tennessee, Knoxville, Tennessee.

Timmer, C.P. and Jayne, T. (2011) *Managing food price volatility: approaches at the global, national and household levels.* Discussion at Stanford University, Stanford, Connecticut. Available at: https://www.youtube.com/watch?v=aHP9e7F11bA

Von Braun, J. and Torrero, M. (2009) *Implementing physical and virtual food reserves to protect the poor and prevent market failure.* Policy brief 10. International Food Policy Research Institute, Washington.

Wright, B. (2009) *International grain reserves and other instruments to address volatility in grain markets.*, Policy Research Working Paper 5028. World Bank, Washington.

Williams, J.C. and Wright, B.D. (1991) *Storage and Commodity Markets.* Cambridge University Press, Cambridge, UK.

World Bank (2008) *Preparing for the next global food price crisis.* Speech by Justin Lin, Senior Vice President and Chief Economist at Roundtable on 17 October 2008. World Bank, Washington.

World Bank (2012) *Using Public Foodgrain Stocks to Enhance Food Security,* Report 71280-GLB. Agriculture and Rural Development, World Bank, Washington.

11 Conclusion

Food and agricultural prices are volatile mainly because sometimes the harvest is good, sometimes the harvest is bad and at other times the harvest is average. When the harvest is good, prices tend to fall. When the harvest is bad the opposite happens: prices tend to rise.

There are many factors that determine the size of the harvest, of which the most important is the weather. Favourable weather means a good harvest. Poor weather, such as a drought or a flood, means a bad harvest. In principle, a bad harvest can also be caused by a lack of labour during a critical point in the growing cycle – such as planting, weeding and harvesting. A lack of labour can be the consequence of a governmental restriction placed on the movement of farm workers, for public health reasons, during an epidemic or pandemic.

Many vicissitudes are visited upon countries by volatile prices of food and agricultural commodities. Hunger, farm bankruptcies, food riots, farmer demonstrations, a lack of agricultural growth – all can have their origins in unstable grain prices. In an effort to avoid these trials and tribulations, this book has put forward two arguments: firstly, that countries gain many benefits when they moderate the volatility of the price of grain on their home market; and secondly, that the most plausible way for a government to address the problem of price volatility is by means of a grain buffer stock.

In principle, the benefits of a more stable grain price include the following: (i) consumers can always afford to buy food; (ii) farmers can always make an acceptable profit from farming and are not reliant on 'public handouts'; (iii) governments do not have to cope with food riots or farmer demonstrations; and (iv) countries are assured of tomorrow's food supply.

A buffer stock is the most plausible way to address the problem of price volatility for the following reasons:

1. It strikes at the root cause (variations in supply and demand) rather than at the symptoms (6.4).
2. Although the government has to finance the capital cost of a buffer stock, once the buffer stock has started to operate, it generates its own

revenue. Unlike consumer food subsidies and farm deficiency payments, a buffer stock is not a drain on the government's budget (9.3).
3. In addition to bringing about a more stable price of grain within the country itself, a buffer stock helps to stabilize the world price of grain. In this sense, a buffer stock is 'globally friendly'. This is in stark contrast to the alternative of trade measures. These cause collateral damage to other countries by destabilizing the world price (8.7).
4. A buffer stock improves economic efficiency by removing the 'noise' from the market price due to random events such as extreme weather (9.5.3).

To a certain extent, a buffer stock can also help to avoid panic buying. If there is a fear that food will, for some reason or other, become scarce some people may rush to the supermarkets and shops. They buy more food than they immediately need because they do not wish to be left without. Panic buying can be a self-fulfilling prophecy because people who come later find that the shelves of the supermarkets and shops are empty and that there is now no food left for them to buy. This was the experience in some countries in the early part of 2020, during the coronavirus pandemic. For public health reasons, governments restricted the movement of people by imposing 'lock-downs'. These had the potential to stop migrant farm labour from working on farms and in the food chain. There was a fear that lock-downs could disrupt, not only national food marketing chains, but also international food chains and lead to food shortages.

Governments can address this fear if the country has a grain buffer stock *in situ*. They can remind their citizens of the existence of a grain buffer stock from which grain can be released in the event of a disruption to the food marketing system. Thus, to some degree, a grain buffer stock may help to calm citizens and to avoid the panic buying of grain-based food products.

Over and above these humanitarian, political and economic issues, there is a strategic issue. The world is facing a number of daunting challenges, of which climate change is one of the most acute. There are several linkages between climate change and farming. First, there is the effect that changing weather brings to farming (more or less rain, higher or lower temperatures and so forth). To be in a position to adapt their farming methods to new weather patterns, farmers need to be assured that the business of farming will be profitable in the future. A precondition for profitable farming is stable farm product prices.

Secondly, there is the effect that farming has on the level of greenhouse gases. Contrary to what some people believe, farming in a number of countries has developed in a way that is harmful to the climate (emissions of CO_2 from machinery and CH_4 from some livestock, for example). Farming needs to use less fossil fuels so that it emits fewer greenhouse gases. Farmers will not be able to adopt low-carbon methods of production if they are having to contend with volatile prices.

There is a third linkage: climate change – like all global challenges – demands a truly global effort. It is perhaps the first endeavour that requires all governments to be 'on board and on the same page'. It is rather pointless for, say, half the world to cut its emissions of greenhouse gases if the other half does not. So long as poor countries remain poor, they will be in a difficult position to make their rightful contribution to the resolution of climate change and other global issues. They will struggle to be part of the global solution. For them to play their full role in the resolution of global issues, poor countries have to be placed on a path towards prosperity. This requires that their farm sectors can develop and grow. One essential condition for agricultural growth – as we noted in the Introduction in our *tour d'horizon* of economic history – is stable farm prices.

In conclusion, when food and commodity prices are stable, we are all spared many difficulties. It is easier to address global problems. Governments need to consider strategies to stabilise commodity prices in which grain buffer stocks play a central role.

Appendix: The International Wheat Agreement – A Case Study

Introduction

The first international agreement to stabilize the price of a commodity was the International Wheat Agreement, signed in 1933. The purpose of this appendix is to look into the background and design of the agreement.

Background

In the 1930s there was a worldwide problem of oversupply of grain. Low grain prices were good for consumers but they spelled doom and gloom for farmers the world over who received a rock-bottom price for their produce. Farmers in countries which imported grain were threatened with losing their livelihood because foreign grain was cheaper. Farmers were angry – in their minds, they had worked honestly and honourably the whole year to produce a good that was essential to society and yet, when they sold their produce, they were offered prices that were derisory. Within Europe, some farmers felt so bitter about their condition that they became supporters of the extremist nationalist parties that were emerging at the time, such as the Nazi Party in Germany, the Front Nationale in France and the fascist party of Italy. Generally speaking, some rural areas were at least sympathetic to these extremist parties and the nationalism they espoused. The growth of these parties was one of the factors in the tensions between European countries which led to World War II. Chaos on the farm was having a political impact. The problem of low farm prices was spilling over into global politics.

To prevent their farmers from going out of business, governments started to place import taxes on foreign grain – in other words, there was a move towards protectionism. This move by some countries away from a global free market prompted governments to try to bring some order to the world market so that prices did not plummet and put their farmers at risk of penury. Twenty-two countries – those that were the world's most important regarding international trade in wheat – came together in 1933. They negotiated the

International Wheat Agreement. The first article stated forthrightly: 'The objectives of this agreement are to assure ... stable prices.'

Effectiveness of the Agreement

At first, the agreement did not work. Governments modified the mechanisms to stabilize prices, and in 1949 came up with an agreement that seemed to be effective. We have already noted, in Chapter 3, the comment by Sewell (1992) that, at the time, it was astonishing that between 1954 and 1968, the price of wheat remained within the narrow range of $8 per ton over 14 years. Compared to the disastrous situation before World War II, this was a great accomplishment.

The 'economic provisions' of the agreement – the practical mechanisms for stabilising the price of wheat – were as follows:

1. The governments agreed a price range for wheat. This consisted of a maximum price and a minimum price. The intention was for all transactions (i.e. all exports and imports) between the participating countries to take place at a price within this range.
2. The governments agreed that they would buy and sell grain with other countries only at prices within this range. They undertook not to buy or sell grain at prices higher or lower than this range.
3. The governments recognized that the world equilibrium price would not necessarily stay within the price band. Thus they agreed on measures to keep it in the price band. These measures were as follows:
4. In years of plentiful harvest, when large quantities of grain would normally be offered onto the market and the world equilibrium price could be expected to fall below the minimum price, the countries agreed to individually withhold a proportion of their own national harvest from the market. This would, it was reasoned, prevent prices falling below the designated minimum price.
5. As for years of bad harvest, when a small quantity of grain would normally be offered onto the market and the world equilibrium price could be expected to rise above the maximum price, the countries agreed to individually sell grain from their own national stocks onto the market. This, it was reasoned, would prevent prices from rising above the designated maximum price.
6. By selling grain in times of shortage and withholding grain in times of surplus, the countries reasoned that the world equilibrium price could be kept within the price band.

In the 1950s, this system worked well. During the sixties, however, it started to come under strain. The grain shortages of the post-World War II period had turned to surpluses. Worldwide production of wheat had been on the increase. In order to keep the world equilibrium price within the designated price band, the major wheat-growing countries – the USA and Canada – had had to withhold ever greater quantities of their harvests from the market. They had had to refrain from selling them, keeping them in store in anticipation of a year of bad harvest when they could have sold them. It was costing them money to store the grain and, by not selling it, they were denying themselves export earnings. Bad harvests, which would have given them opportunities to offload some of their growing surplus stocks, did not arrive. The situation was unsustainable and in 1969, the USA and Canada started to sell grain at less than the designated minimum price. In doing so, they abrogated the agreement. Although the agreement has survived to this day as a forum for the exchange of information, this marked the end of all attempts to stabilize the price of wheat on the world market.

Lessons Learned

First, the agreement did not cover the entire market. Not all countries that traded grain joined the agreement. These non-members were not bound by any rules regarding the price at which they could buy and sell grain to each other. The agreement only ever covered two thirds of the volume of grain that was traded internationally. This meant that there was a parallel market in which the price was sometimes higher and sometimes lower that the designated band. The danger was that if the price outside the agreement was higher, then a country wanting to sell its grain was tempted to infringe the agreement and sell at the higher price. If the price outside the agreement was lower, then a country wanting to buy grain was equally tempted to buy at the lower price in the parallel market. The existence of an alternative market in which prices were higher or lower than the designated band posed a threat to the integrity of the agreement.

Secondly, the price band did not stay in line with the long-term world equilibrium price. During the 1960s, the price band had been raised and the world equilibrium price had started to drop. Sewell (1992) has written:

> The price ranges were raised in the (mis-placed) hope that markets, which had temporarily risen in the mid-1960s after monsoon failures in South Asia, were going to remain tight in the long term. Grain prices started to fall sharply as a result of abundant supplies coupled with a drop in demand, and export prices dropped below the prescribed minima.

The price band was therefore too high relative to the long-term trend in prices. There was no mechanism for tracking the long-term trend of the world equilibrium price. What began as a scheme to stabilize prices was degenerating into a scheme not only to keep prices stable but also to keep them stable at an artificially high level. While the price band remained above the world equilibrium price, this benefitted the selling countries and disadvantaged the buying countries.

Thirdly, there was no mechanism for the costs and benefits of the agreement to be shared fairly between the participating countries. The USA and Canada had been withholding their grain from the market for almost a decade, entirely at their own cost. The rest of the world was the beneficiary of the self-restraint of these two countries. Eventually, the USA decided that the arrangement was not fair. They objected on the grounds that they were paying for a system that benefitted the world as a whole but for which the rest of the world was not paying its fair share.

The fourth flaw was that there was no mechanism by which the rules could be enforced. The agreement was essentially a voluntary agreement. We have already seen that the USA and Canada decided, of their own volition, to disregard the agreement regarding the minimum price. Countries could come and go as they wished. This is what the UK decided to do in 1953 on the grounds that the maximum price was pitched too high. It withdrew from the agreement and did not rejoin until 1959. Since the UK was the world's most important importer of wheat at that time, its departure was a big blow to the viability of the organization.

Conclusion

The International Wheat Agreement was a well-intentioned effort to cope with the age-old volatility of grain prices on the world market. It was successful until the late 1960s. Then it broke down because it was trying to stabilize the price at a level that was higher than the long-term world equilibrium price. It was continuously requiring grain to be kept off the market. The cost of keeping grain off the market was not shared fairly by all the members. The cost fell upon the producing countries rather than on the consuming nations. Canada and the USA were the main producing nations. By the late 1960s they were refusing to pick up the tab on the grounds that other countries were taking advantage of their benevolence. The agreement then lapsed.

This book is not proposing a repetition of the International Wheat Agreement. In effect, the International Wheat Agreement set up a *de facto* international buffer stock that was held and funded by individual nations (in practice, by the USA and Canada), the purpose of which was to keep the world price within a price band. This book is far less ambitious. It is not proposing a world scheme. It is proposing that governments set up their own national buffer stocks to keep the price on the domestic market within a national price band. Furthermore, this book recognizes the need for the price band to always straddle the long-term equilibrium price, which the International Wheat Agreement failed to do.

Reference

Sewell, T. (1992) *The World Grain Trade.* Woodhead-Faulkner, New York and London.

Index

Note: The page numbers in bold, italics and with 'n' represents tables, figures and notes respectively.

Africa, as food deficit continent 12
Asia, as food deficit continent 11
attitudes towards price stabilization 160

biofuels 9, 28–29, 53, 64
buffer stock
  attitudes towards buffer stocks 160
  and economic efficiency 152, 157, 162, 170
  depletion, avoiding *145*
  and compatibility with market 151–152
  and commodity speculation 153, 157
  cost of purchasing grain *143*
  financial costs 141–142
  legal aspects (WTO) 153–155
  limitations of 150–151
  location of 146, *147*
  national/regional 156–157
  operations
    cycle of sale/purchase 148, *148*
    quantity to be purchased 150
    release of grain 148–150, 150
    setting/adjusting the price band 142–146, 158 n6
    storage period 146
  and price neutrality 152–153
  purchase of grain and effect on equilibrium price *87*
  regional (multi-country) 2, 3, 156–157
  revenue generated by sale of grain *144*
  sale of grain and effect on equilibrium price *86*
  sustainability, financial 2, 141–142, 161
  transparency 155
  timing of release of grain *150*
  unwanted farm surpluses (US, EU) 153, 162
  virtual buffer stock 167 n3
  world buffer stock 3

Canada
  and international wheat agreement 3, 174
China
  and 2007/8 world food crisis 140, *141*, 157 n3, 157 n4
  grain buffer stock 2
  as grain importer 13
  slowing rice yields 24–26
climate change 31, 170–171

cocoa, international commodity agreement 3, 161
coffee, international commodity agreement 3, 161
collateral damage
caused by trade measures 37 n10, 119, 133, 139 n5
commercial farmers
definition of 63
commodity speculation 30, 31, 167 n3
consumer
definition of 64
consumer subsidy 90–92, *91*
consumer surplus 163–164
cooperation, international 20, 159 n9
Corn Laws, in Great Britain 4–5
coronavirus 170
crop insurance 96 n1
crop yields *25*, *26*
cropland, world distribution of *13*

deflation 35 n1
deficiency payments to farmers 93, *94*, 99 n11
development, role of agricultural prices in 4
demand
contraction of 43–44, *44*, **44**, 49, 57, *58*, 59, *60*, 63, *75*
derived 51, *52*, 62 n2
expansion of 35, 43, *44*, **44**, 49, 57, *58*, *59*, 63, *69*, 70, 98 n6, 132
for grain 51–53
for imports 101, **102**, *103*, 105–118, 120–125, 133–135
price elasticity of 45, *48*, 49, 62n1
primary 51–52, *52*
schedule 41, **41**, 91, 102, 110
shifts in 41, 43, *44*, 49, 56, 59, 56–61, 83
dollar (US), value of and effect on grain price 27

economic efficiency 152, 170
economic equilibrium 42, *42*, 49
economics of agriculture 68
economics of buffer stock 85
economy
closed 81
open 106, 119
England
agricultural price policy 4, 6
first country to industrialize 8 n3
Europe, as food surplus continent 12
European Union
absence of buffer stock 2
sovereignty sharing 3
export
bans and restrictions 9, 29, 38 n21, 124, 138 n3
commodities 8 n2
subsidy 20, 128, 130–133, *131*, 135–139, *136*
tax 127, 128, 130, 138, 138 n1
exporting country 13, 124–127, 137

famine(s) 6
FAO food price index 16
farm surpluses 153, 162
Farmer Pete 72, *73*, 79 n3
farmers, definition of 63
farmers' toast to next drought 68
fixed price 81, 95–96, 99 n14
'food availability decline' explanation 6
food banks 66, 98 n10, 99 n11
food, expenditure on 66
food mountains 2
food rationing 92–93, 150
food riots 1, 16, 66, 164
food security 7, 155
food stamps (United States) 66, 77, 90, 99 n11
food subsidies 90, *91*, 170
France 24, 172
agricultural price policy 5, 6
free market 26, 83, 95, 162, 165, 172
free trade 1, 4, 7 n1, 101, 127, 162, 165
French revolution and high food prices 5, 66
Friedman, Milton 162
futures contracts 30–31, 96 n1, 167
use by farmers of 96 n1
futures markets 30–31

Galtier, Frank 4
General Agreement on Tariffs and Trade (GATT) 7 n1
Germany
  agricultural price policy 5, 6
global warming 31, 170–171
'globally friendly' stabilization measures 119, 132–133
government budget 170
grain
  exports **14**
  imports **14**
  importance/uses **10**
  maize production, distribution of *11*
  rice production, distribution of *11*
  surplus/deficit continents 11–13
  trade 13–15, *15*
  wheat production, distribution of *12*
  world cropland, distribution of *13*
grain prices
  FAO food price index 16–17, *17*
  and global warming 31, 33
  longer-term supply/demand equation 34
  mid-1970s 22
  price cycles (1996–2008) 22–23, *23*
  1990s 17, *18*
'Great Grain Robbery' 21
green box measures (WTO) 154

harvests, variable and effect of buffer stock *88, 114, 115*
hedging of price risk 30, 96 n1
hoarding of grain 156
hunger 16, 66, 69, 76

import
  tax 128, 131, 132, 138, 138 n1, 172
  subsidy 128, *129*, 130, 133–139, *134*
importing country 13, 28, 123, 125, 127
India 27
  buffer stock 2, 158 n8
  rice export ban 29
inflation, caused by food shortage 35 n1
international commodity agreements 160
International Food Policy Research Institute (IFPRI) 167 n3
International Monetary Fund (IMF) 20
International Wheat Agreement 3, 20, 31, 160
  background 172–173
  effectiveness 173–174
  flaws 174–175
Ireland, farm protest *74*

Japan
  agricultural price policy 5, 6

Koning, Niek 79

Law of Comparative Advantage 101
loss of interest 161

market efficiency 2, 26, 151
market, model of
  conventional supply and demand model 41–51
  international grain market model 100–118
  modification of conventional model 51–62, *55*
market power 71
marketing board 95–96
  and parallel sales 99 n3
marketing chain 64, **65**, 66, **67**, 71
marketing margin analysis 52, 62 n2
monopoly power 95
monopsony power 95

neo-liberal ideology 162
Newbery and Stiglitz 162, 164
'noise' in the price 152

oil crisis 21
oil prices
  and food price rises 27
oil reserves 2
options on futures contracts 97 n1
Organisation for Economic Co-operation and Development (OECD) 20
Organization of the Petroleum Exporting Countries (OPEC) 21

pandemic xiii, 56, 170
panic buying 98 n6
paradox of plenty 72, *73*
price
  ceiling price 35
  consumer price 5, 94
  domestic equilibrium price 81–95, 100–110, 128 n3, 142–149
  fixed price 81, 95, 99 n14
  floor price 35, 99 n14, 153
  long-term equilibrium price 142, 145, 146, 152, 172, 175
  market price 40, 45, 61, 81, 83
  maximum (Pmax) and minimum price (Pmin) 82, 83
  parity price (in US) 97 n3
  world price 2, 4, 16, 28–31, 100–118, 119–139
price band 81, *82*, 82–83, 100, 119
  setting and adjusting 142, 158 n6, 174
price cycles 22–28
price elasticity
  demand
    definition 46
    perfectly elastic 47, *48*
    perfectly inelastic 47, *48*
  supply
    definition 49
    perfectly elastic 49, *50*
    perfectly inelastic 49, *50*
price neutrality 152
price signals 2, 61, 85, 96, 149, 151
price volatility 31
  causes of 40–62, 169
  conditions associated with food price rise **32**
  consequences of fall in price 70–72, 75–76
  consequences of rise in price 65, 66, 68–70
  human consequences of 76, **76**
  mitigation measures 95
  principle of mitigation 89–90
processing costs 64
producer surplus 163–164
production quota 98 n8

regional organizations 2, 3, 140, 156–157
regional stockpiling system 159 n9
revenue, generated by buffer stock 142, 170
Ricardo, David 5, 101, 127
rubber, international commodity agreement 3, 161
Russia 13, 26, 29
  agricultural price policy 5, 6

Sen, Amartya 6, 79 n1, 161
small country and world price *117*
Smith, Adam 61, 162
  and grain shortages/surpluses 1
sovereignty, sharing of 3
Soviet Union, agricultural prices and collapse of 5
speculation, commodity 30, 153
starvation 66, 69, 70, 72, 92
state responsibility for price stabilization 96 n1
Steinbeck, John 73
stockout 89, 149, 150, 153
stocks
  level of world stocks 26, 31, *33*, 37 n11, 133, 166
stocks-to-use ratio 26, *33*, 37 n11
storage 146, 155
  private/public 155–156
  seasonal 155–156
  multi-annual 155–156
strategic petroleum reserve 2

subsidy
  consumer food subsidy 90, *91*, 170
  import subsidy 128, *129*, 130, 133–139, *134*
  export subsidy 20, 128, 130–133, *131*, 135–139, *136*
subsistence farmers 63
sugar, international commodity agreement 3, 161
suicide, farmers 75
supply
  contraction of 29, 45, **45**, *46*, 49, 55, *57*, *59*, 60, 63, 65–69, *65*
  expansion of 33, 45, **45**, *46*, 55, *56*, 59, *60*, 63, *70*, 71, 85
  of exports 100–118, **102**, *103*
  of grain 53–54, 62 n3
  price elasticity of 45, 49, *50*, 140
  quota 98 n8
  schedule of 41, **42**
  shifts in 45, *46*, 54–56, 57–61, *84*
surplus stocks (US, EU) 153, 162

tax
  export 127, 128, 130, 138, 138 n1
  import 128, 131, 132, 138, 138 n1, 138 n2, 172
Thailand, agricultural price policy 6
theory of rent 5
time period (short-term/long-term) 51
Timmer, Peter 4, 79, 166
tin, international commodity agreement 3, 160
trade
  benefits of 100
  between two countries **104**, *105*
  between four countries **108**, *108*
  distortion of 154
  measures 128, 138, 142, 154
  and price stability/instability 120, *121*, 123, *124*, *126*
trade policy
  'globally friendly' 2, 37 n10, 132, 137

USA
  absence of buffer stock 2
  food stamp program 66, 77, 90, 99 n11
  and international wheat agreement 3, 20–21, 31, 160, 172–175

virtual reserves 167 n3

weather, effects on crop yields 29, 31, 55–56, 169–170
Williams and Wright 156, 164
world agricultural output, slowdown of 24
world agricultural trade 7 n1
World Bank 138 n3, 166
world food crisis 2007/8 140
World Food Program 98 n7
world population, distribution of *12*
World Trade Organization 7 n1, 153, 155, 158 n8
Wright, Brian 167